Careers in Aerospace: A Student's Guide to the Sky and Beyond

Careers in Aerospace: A Student's Guide to the Sky and Beyond

Ali Baghchehsara

400 Commonwealth Drive
Warrendale, PA 15096-0001 USA
E-mail: CustomerService@sae.org
Phone: 877-606-7323 (inside USA and Canada)
             724-776-4970 (outside USA)
Fax: 724-776-0790

Library of Congress Catalog Number 2025934581
http://dx.doi.org/10.4271/9781468607116

Information contained in this work has been obtained by SAE International from sources believed to be reliable. However, neither SAE International nor its authors guarantee the accuracy or completeness of any information published herein and neither SAE International nor its authors shall be responsible for any errors, omissions, or damages arising out of use of this information. This work is published with the understanding that SAE International and its authors are supplying information but are not attempting to render engineering or other professional services. If such services are required, the assistance of an appropriate professional should be sought.

ISBN-Print   978-1-4686-0710-9
ISBN-PDF   978-1-4686-0711-6
ISBN-epub   978-1-4686-0712-3

To purchase bulk quantities, please contact: SAE Customer Service

E-mail: CustomerService@sae.org
Phone: 877-606-7323 (inside USA and Canada)
             724-776-4970 (outside USA)
Fax: 724-776-0790

Visit the SAE International Bookstore at books.sae.org

Publisher
Sherry Dickinson Nigam

Product Manager
Amanda Zeidan

**Production and
Manufacturing Associate**
Michelle Silberman

This book is humbly dedicated to the thousands of young students and aerospace engineering enthusiasts, who are on a mission to learn about aerospace engineering in their early careers, but lack a good source of guidance and struggle to find answers to their questions or even realize how much they truly love aerospace, just as I didn't at first. I hope this book speaks to them.

This book will help you—don't worry. I have felt the same frustration of knowing I loved aerospace but not having a unique, reliable resource to guide me. As a result, I stumbled through different paths to discover my passion for aerospace.

In this book, we will teach aerospace-related concepts in the simplest way possible, with an engineering perspective, to help you discover your exact direction in aerospace engineering.

Welcome to Aerospace Engineering!

Contents

Chapter 03 - Spacecraft Propulsion

Chapter 04 - Space Exploration

Chapter 05 - Orientation and Control

Chapter 06 - Communication with Spacecraft

Chapter 07 – Career Overview

Chapter 08 – Startups: Revolutionizing Aerospace Engineering

Author's Note

As a young aerospace engineering enthusiast, I started reading a book about engines to gradually acquire knowledge and skills about the mechanical portions of aerospace and aeronautical vehicles, and I was confident enough to read and understand technical engineering books, but I encountered many difficulties as well.

The very first problem I faced involved the drag force that I saw depicted in aerodynamic illustrations. The first chapter of the book was about differential definitions of force components. At the beginning, I had no knowledge of differential equations and I had never heard of Reynolds. Therefore, I started reading about differential equations, which we can describe as a set of operations on a function. But what are those symbols describing? Well, a function consists of one or more variables. Oh wait, I only wanted to know about "thrust," which is a reversed force with respect to our moving object, because there is friction while something is flying, exactly like the force resisting motion on a solid surface.

Meanwhile as an accomplished aerospace engineer, I still felt the lack of early career educational material, particularly in aerospace engineering, so I have tried to add this book to library shelves to help aspiring students. In this book, I have covered concepts which are not easy to understand, but I have tried to keep things clear and simple and as long as possible without relying too much on math. I have tried to include the information required to give you a good start and enough to enable you to move forward on your own, grow, and help this inspiring field grow.

We definitely will need to hear back from you to improve this book and make it a better source for the next generation of young engineers. Keep in touch with us and tell us how to improve this book through ali@plasmos.space—I would love to hear from you. If you liked it or not or noticed any problem with materials or the copyright or permissions, please do not hesitate to email me and I will definitely review and resolve the problems. It might be possible that we missed granting permission for reuse somewhere, although we did our best to double check that everything, if copied, has proper permission.

This book is intended to help students like me who are just starting their careers. I wanted to ease their path by teaching them some of the things I struggled with early on. Therefore, as long as I am alive, I will do my best to keep the price of this book low. I hope you enjoy it.

Best Regards,
Ali Baghchehsara

Acknowledgments

The authors would like to thank the persons and organizations listed below for permission to reproduce material for some of the figures and materials in this book, and for their assistance in other ways.

Dr. Edward Gibson (NASA Skylab III), Carol Haynes, Ramtin Jamshidi, Dr. John D. Anderson (Scientific Institution), Wikipedia Foundation, Ronald Muller (NASA Goddard Space Center), Dr. Andrew Hunter (American University), the National Aeronautics and Space Administration (NASA), McGraw Hill Education, Science Buddies Organization, and NASA Glenn Research Center (GRC), for providing helpful material. Mr. Emmaneul Carreto, Royal Aeronautical Society, Jens Strahmann, and Barry P. Shaffer for their powerful editorial support.

Thanks to Danielle Nuetzel for helping with the editing of the new edition, and also shoutout to Nicholas A. Vigneron, a future aerospace engineering student, for their feedback used in the second edition to make this book a better resource.

Foreword

By:

Edward G. Gibson, Ph.D.

Science Pilot, Skylab III

It's an Exciting Time to Be Alive!

We humans have a fundamental urge to explore. Whether it be around the next corner, over the next hill, into the air, or out into space, we all have that urge. To different degrees, we each are driven to explore and utilize what we discover.

Aerospace engineering is the enabling technology for flight just above Earth or out into unbounded space far from Earth. This is the topic of this book.

Although humans have jealously admired the capabilities of birds for thousands of years, the application of engineering principles that enable powered flight is just over 100 years old. These applications started with the 12-sec, 120-ft flight of the Wright Brothers in 1903. Innovation and advancements have been rapidly created. In air-breathing aircraft, we have flown 3.3 times the speed of sound at heights over 123,000 ft (37 km). Rocket-propelled aircraft have greatly surpassed these achievements right up to the performance of a space shuttle, which blurs the boundary between aircraft and space flight. Humans have been to our moon and robots out to the edge of our solar system. Clearly, the urge to explore, to utilize, and to push our boundaries is extremely compelling.

Aerospace engineering builds on several basic disciplines including mathematics, physics, chemistry, mechanics, electronics, and communications. Even just a rudimentary understanding of these fields enables a more rapid and appreciative understanding of the advancements in aerospace engineering. Whether you are an interested spectator or a professional in the field understanding the principles of aerospace engineering will expand your wealth of knowledge. Over the years, aerospace progress has demanded the further development of existing technical fields and the creation of new ones based on the above basic disciplines. They include materials science and solid mechanics, structural and fluid mechanics, electronics and software, propulsion and control engineering, noise generation and control, risk and reliability, orbital mechanics, and flight test. Obviously, a wide variety of ways to contribute and enjoy aerospace advancements exist.

Our real limits, even our self-imposed ones, are far beyond our current perception and will challenge us for many centuries to come. In aviation, we continuously seek higher aircraft speeds at higher altitudes coupled with higher reliability and lower costs. In space flight, we also seek lower costs and higher reliabilities coupled with ever greater reaches outward from our planet. On my Skylab mission, we traveled 35-million mi in 84 days. We all thought it was quite an accomplishment. But in a moment of reflection upon landing, I realized that this distance is covered by light in only about 3 min. Yet it takes light over four years just to reach our nearest star. Thus, when it comes to real space travel, we have barely nudged the tip of our collective toe out the front door.

Truly, in air and space, we are on the front end of something much larger than any of us can now imagine. Travels and adventures far greater than anything we can now picture. The fundamental disciplines of aerospace engineering must and will continue their dynamic growth to enable continuous and ever greater accomplishments.

It's an exciting time to be alive—enjoy!

Edward G. Gibson, Ph.D.
Science Pilot, Skylab III
September 2014
Scottsdale, Arizona, US

.

Chapter 01

Flight and Aerodynamics

Figure 1.1 The "Icon-II" future aircraft design concept for supersonic flight over land comes from the team led by The Boeing Company/NASA.

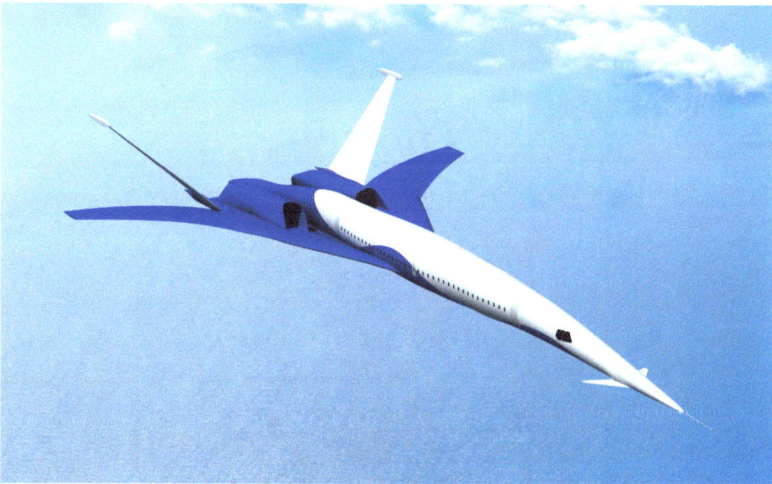

NASA/The Boeing Company.

People study how air passes over and under an object to measure what is called lift, which is the basis for flight. In addition, thrust is required for propelling objects through the air. Lift and thrust enable all that fly (airplanes, rockets, and birds) to defeat gravity and fly. The air, however, resists any object's movement, which is called drag. Lift, force, thrust, and drag as well as the resultant movement are called aerodynamics.

In this section, we will explore how lift and drag work at subsonic speeds (i.e., below the speed of sound). In subsequent sections, we will explore them at supersonic speeds (above the speed of sound).

Liftoff

As an airplane passes through the air, its wings cause changes in the air's speed and pressure, which produces an upward force called lift. Lift is based on what air does in particular circumstances. An aircraft flies because the air supports it. In this chapter, we will explain how this happens.

Birds have been flying for eons, but people have been flying only for a few decades. Humans started with kites and gliders and moved on to create airships, helicopters, planes, and supersonic aircraft. Supersonic flight means flying faster than sound, breaking the sound barrier at 1235 km/h (768 mph).

But how did flying originate? Brothers Orville and Wilbur Wright pioneered it in the United States (US) and were the first to fly an airplane satisfactorily using their newly discovered three-axis control (**Figure 1.2**). Their three-axis control allowed the pilot to direct the plane's flight more accurately.

Though enthusiasts worldwide were simultaneously trying to solve the riddle of powered flight, the Wright Brothers are generally acknowledged as the first to control and sustain it on December 17, 1903. Did you know that plane wings have an airfoil shape? This helps counter the effect of gravity. You might wonder what an airfoil is, but it will be explained in the next few pages.

Figure 1.2 First successful flight of the Wright Flyer, by the Wright brothers. It traveled 36.6 m (120 ft) in 12 sec in a North Carolina photo by John T. Daniels.

Reprinted from the Papers of Wilbur and Orville Wright.

You might already have an idea that the wing creates lift (you will learn exactly what that means later) while it moves through the air. It can be explained by Bernoulli's principle: the airfoil shape makes the air flow more quickly over the wing top than under the wing bottom. This generates higher air pressure below the wing which pushes the airplane upward into the lower air pressure above.

Disturbed air and friction generate drag on the moving plane, thus slowing it down. An engine provides thrust to move the plane forward fast enough to overcome drag and allow the wings to create the required lift.

You might wonder about airships. Do you know what they are? Airships and blimps are buoyant in air, like boats are in water, as they are filled with a gas (often helium) that weighs less than the air around them. The Zeppelin airship was built by Germans (see **Figure 1.3**). A Zeppelin is a rigid airship named after the German Count Ferdinand von Zeppelin who pioneered rigid airship development at the beginning of the 20th century. This design was first patented in Germany and then the US.

Zeppelins were first flown commercially in 1910 by Deutsche Luftschiffahrts-AG (DELAG). DELAG established a scheduled daily service between Berlin, Munich, and Friedrichshafen in 1919.

Figure 1.3 The Zeppelin airship in April 1913 sailed from Friedrichshafen and shortly after ran into bad weather and landed voluntarily.

Photo by Jean-Pierre Lauwers.

What about Natural Flying?

Flying and gliding animals (volant animals) have evolved separately many times, without any single ancestor. The only creatures that can fly are insects, birds, and bats. The only mammals that can maintain continuous flight on one level are bats. Flying fish (see **Figure 1.4**) can glide hundreds of meters because they have enlarged winglike fins.

Figure 1.4 A flying fish or Exocoetidae.

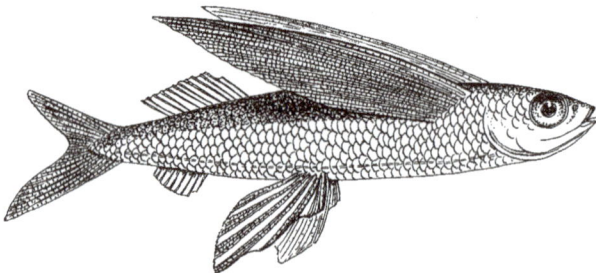

Illustration by Pearson Scott Foresman.

Let us start learning more about aerodynamics and powered flight with some more facts:

- Airplanes are used for transportation, recreation, research, and military purposes.
- An aircraft body is long and thin and is called a fuselage. Basically, fuselage is the main body of an aircraft (without wings).
- Pilots—when required—normally control an aircraft from its cockpit at its front end.
- Some military and experimental aircraft can fly supersonically.
- The Concorde was the most well-known commercial supersonic aircraft (**Figure 1.5**). It flew between London, New York, and Paris until 2003. Its New York–Paris flight more than halved the normal commercial time of flight by roughly 8 h.
- Unmanned aircraft controlled remotely or by computers are known as drones.

Figure 1.5 A touchdown of the Concorde passenger jet. The famous Black Sheds is just visible through the heat haze from the four Olympus engines.

Forces of Flight

If you want to comprehend how flight occurs, you need to learn about the four forces involved: lift, weight, thrust, and drag.

During flight, the wing provides enough lift for the airplane's weight, and the engine produces sufficient thrust to counter the drag and propel the aircraft forward. Increasing the aircraft's weight increases the lift required.

A larger wing might allow more lift, but it would increase the drag and therefore increase the required thrust. These four forces are interdependent: changing one affects the other three (**Figure 1.6**).

- **Thrust** is what moves the airplane forward. It comes from an engine, which can turn a propeller or use jet propulsion. Either method is fine, so long as air keeps going over the wings.

- **Drag** is the frictional force created from the airplane continually bumping into the particles in air. If you walk into a strong wind, you can experience drag resisting against your body. The wind against your body reduces your speed just as the wind against the airplane reduces the airplane's speed. Though airplanes are designed to let air pass around them with the least possible drag, the drag still negatively impacts their thrust.

Figure 1.6 Forces of flight on an aircraft.

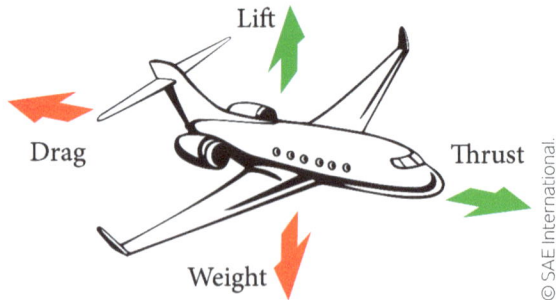

© SAE International.

The minimum lift force an aircraft requires has to be the same as the aircraft's weight. This draws on Newton's third law: "When one body exerts a force on a second body, the second body simultaneously exerts a force equal in magnitude and opposite in direction on the first body." That is exactly what happens in an aircraft to make it fly. To maintain steady altitude in flight:

$$Lift = Weight$$

The effect of an airplane's **weight** is to pull it downward toward the ground. The weight also needs to be balanced evenly from front to back for the plane's safety. When all four forces support each other properly, an airplane can fly. However, even so, most still need a pilot to fly them!

Airfoils

An airfoil-shaped wing (blade) is shown in the cross-section in **Figure 1.7**. How do airfoils produce lift? The lift relates to the airfoil's shape. Two air particles move at the same instant from the airfoil's leading edge, one going over and the other going under the foil. The air particles will then reach the trailing edge of the foil at the same instant.

Figure 1.7 Airfoil cross-section and airflow.

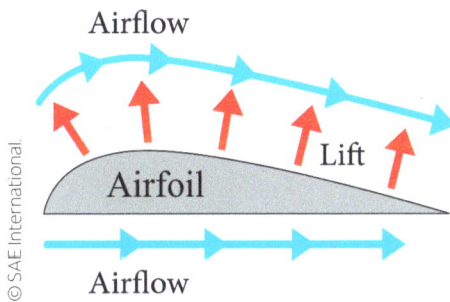

When an airfoil is curved on the top, the air particles flowing over it have to travel further than the ones flowing under it. This forces the air particles going over the top of the wing to speed up so they will meet their counterparts going underneath. The difference in speed creates a pressure differential: the slower air underneath pushes up and the faster air on top pushes down on the foil. As a result, the wing rises higher into the air. This is illustrated in **Figure 1.8**.

Figure 1.8 demonstrates what a steady airflow would look like in a low-speed wind tunnel. Smoke in the upstream reveals the streamlines around the airfoil model. If you search for wind tunnel test pictures on your search engine, many great examples will appear.

Figure 1.8 Lift and airfoil airflow.

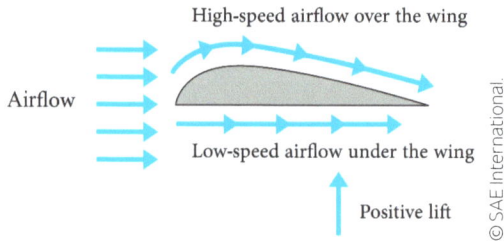

High-speed airflow over the wing

Airflow

Low-speed airflow under the wing

Positive lift

© SAE International.

Physical Definition of a Flowing Gas

In order to fly, an airplane needs a sustained lifting force from the flow of air over its surface. Different surface shapes create different airflows, so airplanes are specially designed to induce an airflow that most efficiently produces the required lifting force. Of course, lift is not the only determinant in airplane design. Any final design is a compromise between multiple requirements. As you progress through this book, you will gain a greater appreciation of those requirements. For now, we will focus on airflow.

Aerodynamics involves the study of flowing gas. Since air is gaseous, aerodynamics is commonly understood as being the study of airflow. A person who studies aerodynamics is an aerodynamicist. This scientific field is fundamental not only to airplane design, but also to propellers, wind tunnels, jets, rocket engine designs, projectile configurations, and vehicles that are to enter another planet's atmosphere from space.

Aerodynamics also embraces the intricate flow of our Earth's atmosphere and the flow of smoke etc., through smokestacks, for example. Since aerodynamic principles apply anytime a gas moves or something passes through the air, the reach of aerodynamics is practically limitless. For you to understand aerodynamic applications, we will discuss the four foundational aspects of aerodynamics: pressure, density, temperature, and velocity.

Pressure

Pressure is perhaps most easily understood through an ordinary everyday situation. Imagine sitting in a moving car and holding your hand out the window with your palm facing forward. Your palm is like a wall—perpendicular to the oncoming stream of air—and you will feel the air pushing your hand backward.

This is air pressure pushing in the airflow's direction. More scientifically, this pressure is the force per unit area on your palm produced by air molecules striking your hand's surface and transferring some of their momentum to it.

The specific definition of pressure is as follows: the normal force per unit area exerted on a surface from the time rate of change of momentum of the gas molecules impacting on that surface. Pressure is usually measured at a certain point in a gas or on a surface. It can vary from point to point, as it might not be uniform either across a surface or in an airflow.

Density

We generally define the density of a gas as mass divided by the volume of its container. In aerodynamics, the density of gas is defined more specifically as the mass per unit volume, where that unit volume can be shrunk to near zero at a specific point. Density, like pressure, might vary from point to point within a gas. In addition, the density of a gas alters when temperature and pressure change. As pressure increases, density increases; as temperature increases, density (most often) decreases. Density also varies with altitude: as altitude increases, air density decreases. This happens because the force of gravity on air molecules higher up weakens, which enables the air molecules to drift further apart.

Temperature

The temperature of a gas is a measure of hotness or coldness. For gases, it is directly proportional to the average kinetic energy of the particles from the gas. So, what does this mean?

Gas consists of molecules and atoms in a state of constant, random motion three dimensionally. The molecules constantly collide with each other and their container. Moving particles always possess both speed and kinetic energy: a fast-moving particle has more kinetic energy and a higher temperature than a slow-moving particle. To find a meaningful measure of the average kinetic energy, we can observe a single particle's numerous collisions over a period of time. Temperature is a point of measure. The temperature may vary from point to point in a gas. A theoretical frozen particle that is not moving at all and therefore has no kinetic energy would thus have a temperature of what is called absolute 0 (which is −273.15°C or −459.67°F).

Velocity

Velocity relates to speed. Speed is commonly understood as the distance traveled by an object per unit of time, as in driving a car at 100 mph. If we mention the direction the plane will travel in, as in 100 mph due south on a horizontal road, we are referring to velocity. Velocity is also known as speed combined with direction. For a flowing gas, though, the speed and/or direction of individual particles might vary. Therefore, the flow of velocity might vary from point to point in a gas particle.

Like pressure, density, and temperature, flow velocity is a point property. But velocity alone is a vector quantity, i.e., related to speed and direction. The two parts of **Figure 1.9** will help you picture flow velocity more clearly. Consider gas flowing through a rocket engine and air flowing over an airfoil.

Figure 1.9 Flow velocity.

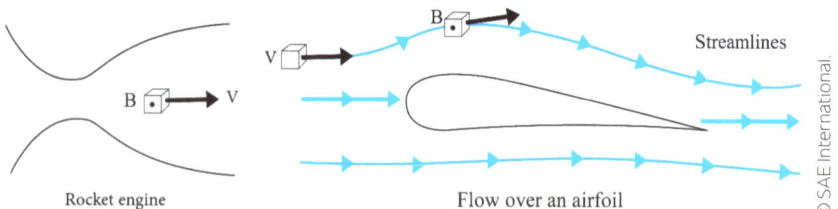

Rocket engine

Flow over an airfoil

Streamlines

© SAE International.

In each case, imagine fixing your eyes on a specific particle and watching them move from point to point in the flow. You will see that the speed and direction can vary. Now focus on point B in each part of **Figure 1.8** and imagine your specific particle passing through it. In a flowing gas, the velocity at any fixed point—both points B in this case—is the velocity of an infinitely minute particle as it travels through B.

Steady Flow and Streamlines

Flow field visualization is a powerful aerodynamic aid. Typically, the flow field shows the paths that gas particles follow. We will learn about **steady flow**, the fluid in motion which does not vary over time—rather, it is fixed. We will also consider a moving gas particle that follows a fixed path in the flow, called a **streamline**.

One method for showing a gas flow is to draw the flow field's streamlines around an object, such as an airfoil. **Figure 1.9** shows the flow's directionality with sketch streamlines. In practice, aerodynamicists use wind tunnels, so they can visualize streamlines in real time. We will discuss wind tunnels in more detail later, but for now, we will focus on some resulting images.

Figure 1.10 shows a parabolic curve in front of the fin from the bow shock wave and flow separation ahead of the fin. Flow is from right to left. The Mach number in this simulation is 5.

Figure 1.10 A surface streamline pattern in supersonic flow.

Compressible and Incompressible Flow

Aerodynamically, there are big differences between flows that are compressible and flows that are incompressible—as you can see in **Figure 1.11**.

Figure 1.11 Compressibility.

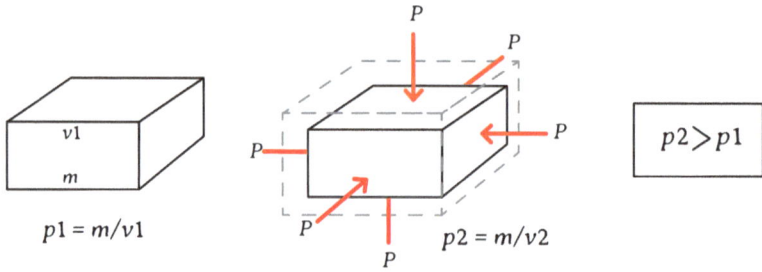

$$p1 = m/v1$$

$$p2 = m/v2$$

$$p2 > p1$$

© SAE International.

Imagine what a trash compactor achieves as it applies pressure to some trash: the trash's mass is unaltered but its volume decreases. In other words, the trash is compressible because its density can change—it can increase.

Figure 1.11 shows a substance being compressed. Its density p alters by an amount that depends on its nature and how much pressure is applied. For water or a solid such as steel, the alteration in volume is very small; so, even though $p_2 > p_1$ in practice, we can consider p to be constant. On the other hand, the volume for a gas can easily change and we can consider p as a variable. Now consider that the material—a gas—is moving along a streamline in a flow.

In a compressible flow, the density of the gas might alter from one point to another, so that p_1 is not equal to p_2. Strictly speaking, all flows in real life are compressible. However, in some cases, the density varies only very slightly and leads to a second type of flow—an incompressible flow. For an incompressible flow, the density of the gas is constant for both time and space. We observe that $p_1 = p_2$ leading us to consider:

$$\text{Area} \times \text{Volume} = \text{Constant}$$

Therefore:

$$A_1 v_1 = A_2 v_2$$

When we think of a material as being incompressible, we might think that its density is unchangeable. Although such a material does not exist in nature, as all matter is at least compressible in some sort of way, the size of the change might be so small as to be irrelevant in a real-world situation.

Aerodynamically, it is useful to treat a gas flow as if it is a solid and classify it as incompressible when its density variation in the flow is so minor. In these cases, we assume that p is constant in order to simplify our analysis of the problem. This type of simplification is commonly used in engineering and physical sciences, but it must be employed very carefully. In other words, it should fit the real-life situation at hand.

At low air speeds, where velocity < 100 mps (i.e., <225 mph), we can assume that an incompressible flow is a close approximation to reality. In the early days of flight, from 1903 to the late 1930s, aircraft flying at these velocities were the norm. Therefore, there is a large body of literature dealing with incompressible flows stemming from the early development of aerodynamics. At higher velocities involved in modern aerodynamics, we deal with compressible flows because the changes in density throughout the flow are greater. The analysis has become more complex because p must be treated as variable.

In analyzing and solving aerodynamics problems, you will almost always need to distinguish between incompressible and compressible flows. Since there are notable quantitative and qualitative differences between them, the results (and accuracy) of your calculations may vary greatly, depending on which type of flow you assume.

Speed of Sound and Aerodynamic Flow Regimes

Knowing the speed of sound in a gas is vital for aerodynamics. The speed of sound is the distance per time unit at which a sound wave travels through a medium. Its speed in a gas depends on which gas it is traveling through, its density and altitude. For instance, sound does not travel at the same speed in Mars's atmosphere as it does in Earth's atmosphere. Within Earth's atmosphere, it travels faster at sea level than at airline-cruising altitudes.

Figure 1.12 illustrates how we can work out a formula for calculating the speed of sound. Imagine the illustration represents an enclosed room where a small blast occurs in a corner. The air in the room is motionless (stagnant), and you are in the middle of the room. The blast generates a sound wave—a thin region of disturbance in the gas—that will pass by you at velocity a in mps (ft/s) etc.

The gas in the room has density ρ, pressure p, and temperature T.

Figure 1.12 Model of a sound wave moving in stagnant gas.

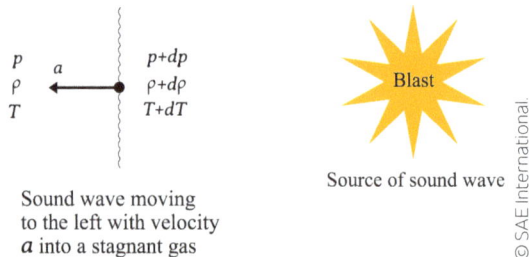

p a $p+dp$
ρ ←• $\rho+d\rho$
T $T+dT$

Sound wave moving
to the left with velocity
a into a stagnant gas

Blast

Source of sound wave

© SAE International.

Using the speed of sound, we can derive a very useful quantity, the Mach number, which is the ratio of the speed of an object to the speed of sound in the surrounding medium. This is an essential quantity in aerodynamics, as fast-moving objects generate high-speed gas flows. Do you recall the illustration in **Figure 1.9**, which considers a fixed point in the gas flow (point B)? The Mach number, M, at point B is the flow velocity, V, at that point divided by the speed of sound, a, at that point:

$$M = \frac{V}{a}$$

This quantity is very useful because with it, we can precisely define three different aerodynamic flow regimes. Each regime has unique characteristics that we will discuss in later sections. Yes, mathematics is sometimes important, but if you bear a little bit, it passes.

- **Subsonic flow:** $M < 1$

  In subsonic flows, the flow velocity in the whole flow is less than the speed of sound.

- **Sonic flow:** $M = 1$

  Mach 1 is the local speed of sound. Do you recall that the speed of sound in air alters with temperature? Consequently, the speed denoted by Mach 1 changes with the altitude.

- **Supersonic flow:** $M > 1$

  In supersonic flows, the flow velocity is more than the speed of sound in the whole flow. When a supersonic aircraft is flying at Mach 2, its speed is twice that of sound.

In addition to these three regimes, there are two more specialized regimes that are loosely defined by Mach number and refer to special sets of aerodynamic phenomena.

- **Transonic flow** contains regions of both subsonic and supersonic flow, where M usually ranges between 0.8 and 1.2.
- **Hypersonic flow** speed greatly exceeds the speed of sound, usually above Mach 5.

Finally, we can see what Mach numbers have to do with compressible and incompressible flows.

Generally speaking, for a low-speed subsonic airflow with a Mach number less than 0.3, only insignificant compressibility occurs. So incompressible flow can be assumed. With Mach numbers above 0.3, compressibility becomes more significant, and its effects should be taken into account. Supersonic, sonic, and some subsonic flows are compressible flows.

Wind Tunnels

We will now turn our discussion toward specific aerodynamic topics of interest, drawing on the fundamentals presented previously in this book. Let us begin with wind tunnels. Wind tunnels are ground-based laboratories that study or experiment with what happens when air moves past solid objects. They are designed to produce airflows that simulate natural flows encountered outside, so scientists can measure the aerodynamic forces affecting the test object. The setup is designed this way, rather than having the test model move at the target speed through the tunnel.

For aerospace engineering, wind tunnels are most often used to simulate the airflows of actual flight conditions experienced by airplanes, missiles, and space vehicles. Since various types of aircraft travel at very different speeds and operate in very different flight conditions, wind tunnel facilities vary in design according to the aerodynamic flow regime they specialize in. For example, from the early days of human flight until around 1940, airplanes traveled at about 480 km/h (200 mph) or slower.

A wind tunnel used to test and develop these aircraft would therefore be designed to produce low-speed subsonic flows. At the other end of the spectrum is the Apollo lunar spacecraft: because it needed to withstand a reentry velocity in excess of 40,000 km/h (25,000 mph), it had to be tested in a tunnel designed to simulate hypersonic flow conditions. Nowadays, there are still plenty of uses for wind tunnels covering subsonic, transonic, supersonic, and hypersonic conditions. The pressure p at various locations in the wind tunnel is related to the flow velocity via Bernoulli's equation, which you can easily find online.

Figure 1.13 illustrates an open-circuit wind tunnel. It contains 13 numbered items, of which we will discuss five. The settling chamber (12) is at the beginning of the wind tunnel. It comprises screens and honeycomb-shaped mesh, for straightening out the air to reduce turbulence.

Figure 1.13 Illustrated layout of a wind tunnel.

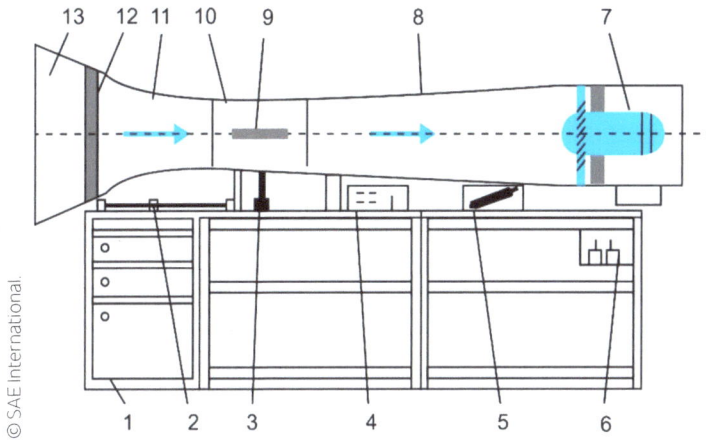

© SAE International

The contraction cone (11 and 13) forces a large volume of air through a small opening in order to increase the wind velocity in the tunnel. The test section (10) is the place where a model to be tested is mounted on sensors. The diffuser (8) is at the end of the test section. It keeps the air running smoothly toward the end by opening wider to slow the air down as it exits the tunnel.

The drive section (7) is at the very end. This houses the fan that faces outward. Its purpose is to draw air into the wind tunnel by blowing air out of it. There are two advantages to using the fan this way. There is less turbulence if the air is drawn out rather than blown in, and the airflow through the tunnel can be more easily controlled. Wind tunnels are not always small. The biggest one is at the NASA Ames Research Center and is shown here. In 1993, the drag chute on the space shuttle required testing (**Figure 1.14a**).

The scientists wanted to improve its drag performance and reduce the buffeting it caused. **Figure 1.14b** shows the inside of the test section when a one-third-scale model of the space shuttle was being readied for tests. It is a 12 by 24-m (40 by 80 ft) test section. This testing resulted in chute modifications that were then used on all subsequent space shuttle missions.

Figure 1.14a The prominent structure in the foreground is the world's largest wind tunnel.

Courtesy of NASA.

Figure 1.14b The test section at the NASA Ames Research Center.

Courtesy of NASA.

Supersonic Flow and Shock Waves

Let us consider what happens if a solid object is placed in a gas flow. The gas molecules will collide with it and create a disturbance in the flow. The disturbance does not stay put, but spreads away from the object to other regions in the flow via weak pressure waves, essentially sound waves, at the local speed of sound. Their impact on the gas flow depends on the flow regime. We will first look at what happens in subsonic flow and contrast that with what happens at supersonic flow to convey the essence of shock waves.

Difference between Subsonic and Supersonic Flow

In a subsonic flow, pressure disturbances travel more quickly than the flow velocity. By that means, they even travel upstream from the object and are felt in all regions of the flow. In a supersonic flow, pressure disturbances travel more slowly than the flow velocity, and as a result they do not travel upstream.

Instead, they collect, join up, and form what is called a shock wave at a finite distance from the object. This happens whenever a solid body is placed in a supersonic stream. **Figure 1.15** shows photographs of super-sonic flow over five aerodynamic shapes. The flow region preceding the shock wave does not experience pressure disturbance, but the flow region following it does; the shock wave is a thin boundary in a super-sonic flow, in which flow properties alter drastically.

It divides the undisturbed flow upstream from the disturbed flow downstream. Generally, shock waves cannot be seen, but they can be shown in a wind tunnel by employing a specially designed optical system called a Schlieren system. This system was used to produce the group of images shown in **Figure 1.15**.

Figure 1.15 Simplified demonstration of: (a) Shock waves on a swept-wing airplane (left) and on a straight-wing airplane (right). (b) Shockwaves on a blunt body (left) and sharp-nosed body (right). (c) Shock waves on a model of the Gemini manned space capsule. These are demonstrations based on the real Schlieren supersonic pictures from NASA. These are simplified to be visible in printing.

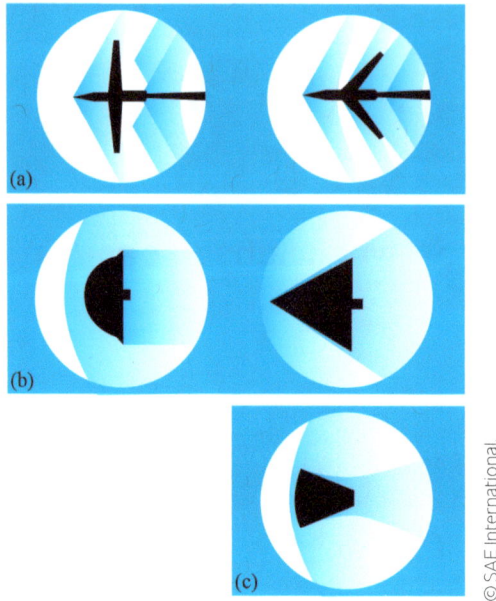

There are a couple of circumstances in which you might see a shock wave. If you are a passenger on a high-speed subsonic airplane such as a Boeing 707 and you look out the window along the wing when the sun is just about straight overhead, you might see shock waves darting about on its surface.

This happens because the airflow over the upper surface of the wings contains localized supersonic regions, which are usually accompanied by weak shock waves. If you are on the ground watching a supersonic jet fly by, you might see a cloud of condensation. This happens because a fast-moving aircraft can compress the air enough to force moisture to condense, producing a visible cloud. When the aircraft passes through the speed of sound, it also sheds water drops that condense on the plane, creating a cone-looking white halo. If you see that, it means that the aircraft has just gone supersonic!

How Does Sound Move (Propagate)?

Vibration begins with mechanical movement, like striking a drum. This causes the molecules adjacent to the mechanical event to vibrate (i.e., the drum skin that was struck). The vibration of these molecules makes the molecules around them vibrate too. This causes the vibration to spread from molecule to molecule, thus causing the sound to travel. Sound can travel only through matter because it needs the vibration of molecules to propagate. Outer space is like a vacuum—it lacks the medium for carrying sound and is therefore very quiet. There is no vibration capable of producing sound in outer space.

Sonic Boom and Quiet Spike: Sonic boom is minimized by using a Quiet Spike. Quiet Spike is an aircraft extension developed by Gulfstream Aerospace. The idea for Quiet Spike is to push some air out of the way before the main body of an aircraft arrives, so the aircraft functions more like a subsonic aircraft. Rather than having a single pressure wave, a more gradual increase in pressure occurs.

What Is the Sound Barrier?

When airplanes exceed the speed of sound (also called Mach 1), they are said to break the sound barrier. Most airplanes do not go this fast, but some fighter jets do. When they do, they create what is called a sonic boom, a very loud explosive noise that is generated by many sound waves being forced together. A sonic boom happens when a plane is traveling faster than sound.

Whenever an object moves through air, it will displace some of that air, which results in a change in pressure (**Figure 1.16**) for waves flowing off an aircraft. Another great example of an object being displaced by air is the effect of leaves being dragged behind a bus or being pushed forward as the bus drives through them. You can sometimes hear when the displacement happens too. When a car passes by a row of parked cars, it may be accompanied by a whooshing sound from the air being moved. The propagation being left behind is sometimes visible when the speed is high enough (**Figure 1.17**).

Figure 1.16 Pressure waves flowing off an aircraft at different speeds.

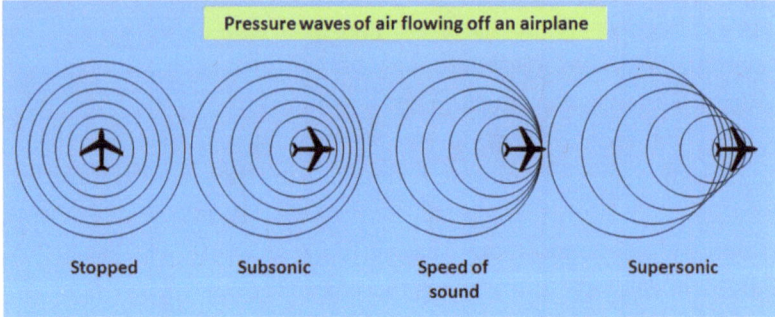

Pressure waves of air flowing off an airplane

Stopped   Subsonic   Speed of   Supersonic
                                    sound

Reprinted from Supersonic: The History of NASA's Sonic Boom Research.

Figure 1.17 An F/A-18 Hornet creating a vapor cone at transonic speed just before reaching the speed of sound.

U.S. Navy photo by Ensign John Gay.

In this picture, the air in front of the aircraft is moving at subsonic speeds and starts getting pushed out of the way by pressure building up in front of the aircraft and being transmitted forward before the aircraft can reach it. This process involves a gradually increasing pressure rather than one single pressure wave. However, air can only respond to pressure changes up to the speed of sound. Beyond that speed, pressure disturbances cannot travel ahead of the aircraft because the air's ability to transmit those changes is limited by the speed at which sound travels—which is itself determined by how quickly pressure changes can move through the air. This is important in aerodynamics because it explains why shock waves form when an aircraft approaches or exceeds the speed of sound—the air can no longer "get out of the way" smoothly, and pressure builds up dramatically.

When one molecule in the air is pushed on, it pushes the adjacent one on too, thus transmitting the change. The speed at which this "pushing on" occurs is dependent on altitude, humidity, and temperature. At sea level, it is about 340 mps (about 770 mph)—or when Mach = 1, but it is less than that at elevated altitudes.

Conclusion

This chapter has provided you with a short introduction to the history and physics of flight. In the next chapter, you will begin to learn about some of the engineering marvels that have revolutionized transportation and our way of life as well as laid the foundation for spaceflight.

Chapter 02

Jet Propulsion

Figure 2.1 General Electric GE turbofan engine with its cowl open. This particular engine is installed at #4 position on the Boeing 747-8l prototype. The CF6 is a large turbofan developing 22,000 to 33,000 kg of thrust. It powers the Airbus A300, A310, and A330, the Boeing 747 and 767, and MD-11.

Jet propulsion revolutionized air travel. Propulsion creates a force that results in movement. A propulsion system includes a mechanical power supply (from an engine, motor, or muscles), and components for generating a force from it (such as a wheel and axles, propellers, a propulsive nozzle, wings, or fins). Some other accessories, like a clutch and a gearbox, might also be necessary for connecting the power supply to the force-generating component. The word *propulsion* comes from the Latin words *pro*, which means "before or forward," and *pellere*, which means "to drive." So they can sustain flight for any appreciable amount of time, aircraft need some type of propulsion system. An aircraft designer has a number of different propulsion systems to choose from. In this chapter, you will learn about jet engines and rocket motors, along with a brief overview of the various propulsion systems available, as well as their advantages and disadvantages.

The purpose of the propulsion system is to produce a controllable force called thrust. Thrust is designed to act together with the other forces impacting an aircraft (lift, drag, and weight) to produce the desired translational motion. In most cases, thrust is used to accelerate an aircraft along its flight path and to counteract drag. However, it is certainly possible for thrust to be used in other ways such as augmenting lift, for example, in a V-STOL aircraft, like the AV-8A Harrier. The combination of a **piston engine** (also known as a reciprocating engine) and a propeller probably provides the most efficient propulsion system for low-speed aircraft.

Figure 2.2 shows a piston engine. Its propellers are adjustable and can increase the momentum of a relatively large amount of air and thus produce positive or negative thrust (when needed). Such engines consume fuel very efficiently too. Normally fuel consumption is measured in kilograms of fuel per hour per brake horsepower and given the name of brake-specific fuel consumption (BSFC).

Figure 2.2 A Rolls-Royce Merlin installed in an Avro York.

Typical values of BSFC required for cruising for the currently used aircraft reciprocating engines are 0.18 to 0.23 kg (0.4 to 0.5 lb) bhp-h. At flight speeds above approximately Mach 0.3 (0.3× the speed of sound). The efficiency of the propeller starts to drop off because of the effects of compressibility on its blades. The engine size necessary to produce the required thrust then makes other propulsion systems more attractive.

A Short History of Early Engines

In the 1700s based on his own third law of motion, Sir Isaac Newton theorized that an explosion directed rearward could propel a machine forward at a great speed. For example, as hot air blasts backward through a nozzle, a plane can move forward. We will come back to this later in Chapter 5. In 1852, Henri Giffard built the first airship powered by a 3-hp steam aircraft engine, but it was too heavy to fly. In 1874, Felix de Temple built a monoplane that flew a short downhill distance using a coal-fired steam engine. In the late 1800s, Otto Daimler developed the first gasoline-powered engine.

In 1894, American Hiram Maxim tried powering his triple biplane with two coal-fired steam engines. It flew for only a few seconds because the engines were too heavy for the flight. American Samuel Langley made model airplanes powered by steam engines. In 1896, Langley successfully flew an unmanned steam-powered airplane called the *Aerodrome*, which ran out of steam after a few kilometers. In 1903, his full-sized gas-powered plane, the *Aerodrome A* crashed immediately after being launched from a houseboat.

As mentioned in the last chapter, the Wright Brothers flew *The Flyer* in 1903, which was powered by a 12 hp gas-powered engine. From that year onward into the late 1930s, only gas-powered reciprocating internal-combustion engines with a propeller were used to propel aircraft.

In 1930 Frank Whittle (**Figure 2.3**), a British pilot, designed and patented the first turbojet engine. The turbojet first flew successfully 11 years later after his design and patent. This engine had a multistage compressor, a combustion chamber, a single-stage turbine and used a nozzle.

Figure 2.3 Frank Whittle with one of his first designed Engines in 1930.

At the same time as Whittle in England, Hans von Ohain was experimenting on a similar design in Germany. In 1939, the world's first turbojet-powered flight occurred in Germany in The Heinkel He 178, which used a gas turbine engine. It was a private venture by the company German Heinkel in accordance with director Ernst Heinkel's emphasis on developing technology for high-speed flight. Heinkel He 178 (**Figure 2.4**) was the world's first turbojet aircraft and jet plane. It was also the first rocket aircraft and flew on August 27, 1939, piloted by Erich Warsitz.

Figure 2.4 Heinkel He 178, the world's first turbojet aircraft and jet plane in Rostock-Laage Airport.

Gryffindor/https://commons.wikimedia.org/wiki/ File.Flughafen_Rostock-Laage1.JPG.

Design of Jet Engines

You have heard of jet aircraft and jet engines; the truth is, modern aviation owes its success to the jet engine. In a jet engine, there are more than a thousand moving parts, but the major components of a turbojet include turbofans, turboprops, and turbo shafts. These are shown in **Figure 2.5**.

Figure 2.5 The component parts of a jet engine.

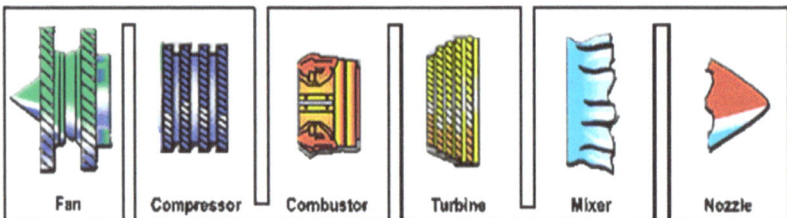

Courtesy of NASA.

Basically, a jet engine is a reaction engine acting in accordance with Newton's third law of motion. The engine sucks in air (from the left to the right) at the front with a fan and discharges a fast-moving jet that generates thrust by jet propulsion. A compressor consisting of many blades is attached to a shaft, which then raises the pressure of the air.

The turbojets blades spin at a high speed to compress air, i.e., squeeze the air. In the combustion chamber called a combustor, the compressed air is sprayed with fuel and an electric spark lights the mixture. On its way toward the nozzle, the hot air passes through a second group of blades, the turbine, which contains a bypass area. Some of the air becomes very hot, and some of it becomes cooler. This cooler air is then mixed with the hot air in the mixer.

Since it is attached to the same shaft, the spinning turbine causes the compressor to spin. As it burns, the gas expands and blasts out backward through a nozzle at the engine's rear, which causes the engine of the aircraft to move forward.

There is another way of looking at this process, which is illustrated in **Figure 2.6**. The fan sucks in the air. The compressor compresses it in the combustion chamber, and then thermal energy is added to the system, increasing the pressure. High-pressured gas makes the turbine spin, as a result making the compressor spin. This is a cyclic system that makes the jet engine operational. Using the energy equation [that no energy is created or dies itself], the jet engine's energy system is:

$$\text{Air}\left[\text{zero energy}\right] + \text{Compressor energy} + \text{Fuel}\left[\text{thermal energy}\right] = \text{Turbine spin} + \text{Thrust}$$

Figure 2.6 The jet engine's energy system.

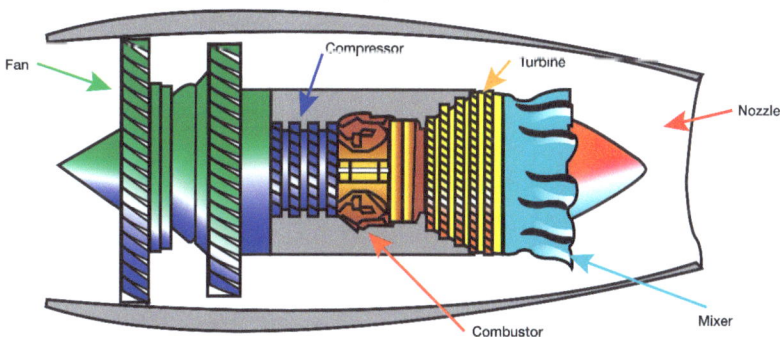

Reprinted from A Survey of Intelligent Control and Health Management Technologies for Aircraft Propulsion Systems. NASA.

Turbofan Engine

The most modern version of a basic gas turbine engine is a turbofan engine. In this, the core engine has two fans: one at the front and one at the rear. Each of them contains many blades just like the core compressor and core turbine.

The engines are connected by a shaft as is the case for the core compressor and the turbine. Some of the engine's fan blades spin with the shaft and some remain stationary. The fan shaft sits inside the core shaft. Such an arrangement is known as a two-spool engine (one for the fan and one for the core). Some advanced engines have additional spools for even higher efficiency. We will look at these parts in more detail now.

Fan: The first part is a large spinning fan sucking in large volumes of air. Most of its blades are made from titanium. The air is sped up and split into two parts. Part of the air passes through the engine's core, as the second engine's components act on the first engine's air. More air travels through the bypass duct surrounding the core toward the rear of the engine, where it produces most of the force that propels the airplane forward. This cooler air not only quietens the engine but also adds thrust to it. The second fan is the one marked Fan in **Figure 2.7**. This and the area marked Fan nozzle indicate the bypass area.

Compressor: The compressor is the first part in the engine core. It is attached to a shaft and has fans with many blades. It squeezes the entering air into progressively smaller areas, which increases the air pressure—and increases the air's energy potential before it is forced into the combustion chamber.

Combustion chamber (combustor): Air and fuel are mixed and then ignited in the combustion chamber. Fuel is sprayed into the airstream by up to 20 nozzles, so that it will catch fire. The fuel and the oxygen in the compressed air burn, producing hot, expanding gases for a high-energy airflow. Internally, ceramic materials are used for the combustor, so it is heat resistant as temperatures reach 2700°F.

Figure 2.7 Schematic of a two-spool turbofan (HP = high pressure and LP = low pressure).

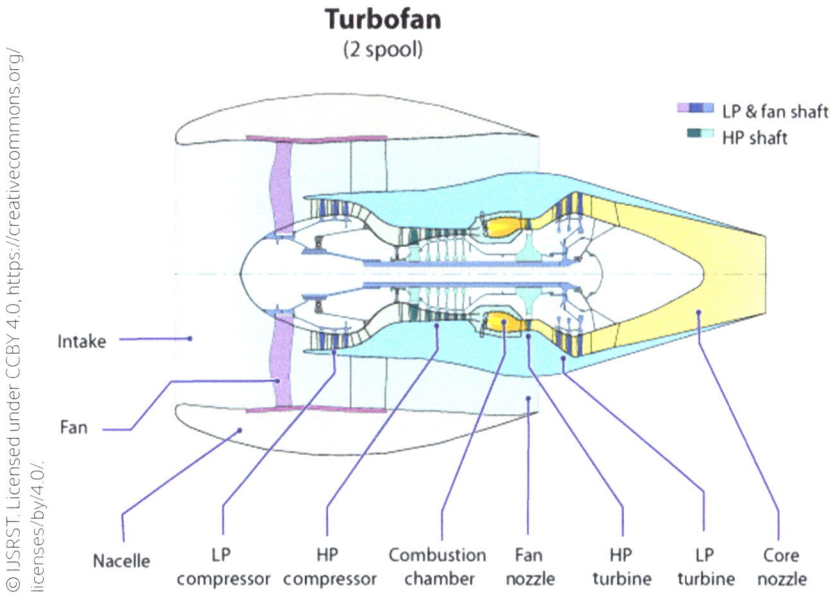

Turbofan: The high-energy airflow then enters the turbine and causes the turbine blades to rotate. These blades are linked by the same shaft that turn the compressor blades and the intake fan. Some of the energy from the high-energy flow is consumed by driving the fan and the compressor. The jet's turbines rotate thousands of times per minute. This is possible because the shafts that they are attached to contain several sets of ball bearings. The core compressor and core turbine are connected to a shaft (and sometimes an additional shaft). Some fan blades rotate with the shaft while other blades remain stationary.

Nozzle: The engine's exhaust duct is called the nozzle. It is the engine part that actually produces most of the plane's thrust. The air exiting the nozzle comes from the hot airflow that has passed through the turbine, plus the colder air that bypassed the engine's core. Together, they produce an exhaust that causes a forward thrust to the engine and therefore to the aircraft. There might be a mixer in front of the nozzle for mixing the hot air and the colder air. This helps quiet the engine.

The turbofan engine provides the most efficient available propulsion for speeds between those of the turboprop and turbojet. The turbofan can be considered a cross between a ducted turboprop and a turbojet, which we will discuss later.

Turbojet Engine

In a turbojet, an opening at the front of the engine takes in air, which is compressed 3 to 12 times more than its original pressure in a compression chamber (**Figure 2.8**). Fuel is then added, and the mixture is burned in a combustion chamber to raise its temperature to 500–600°F.

Figure 2.8 Schematic of a turbojet engine.

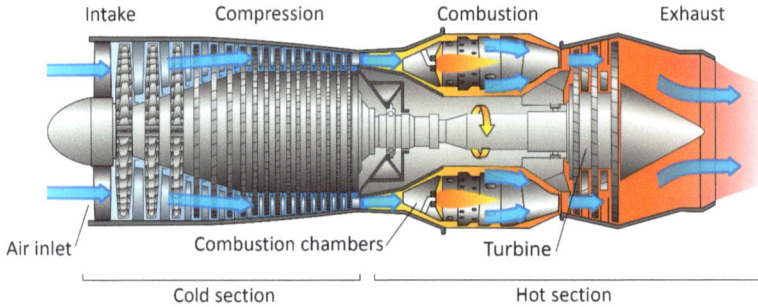

Afterward, the air is fed through a turbine, which drives the compressor. With an efficient turbine and compressor, the pressure at discharge from the turbine is nearly twice the atmospheric pressure. Thrust is produced when the nozzle expels a highly pressurized high-velocity stream of gas, but it can be further increased by using an afterburner. The sections of the engine are similar to the turbofan. It can be said that a turbofan contains a light version of a turbojet and a fan.

Turbojet with Afterburner

Turbojets are usually augmented by afterburners in supersonic high-performance aircraft. The only difference between this and normal jet engines is the addition of a long afterburner duct. After air passes through the turbine, additional fuel can be selectively added and burned in the afterburner duct. This exhaust flame may show shock diamonds, which are caused by shock waves formed due to differences between ambient pressure and exhaust pressure. These imbalances cause oscillations in the exhaust jet diameter over a distance and cause visible banding where the pressure and temperature are highest. You can see these shock waves in **Figure 2.9**.

Figure 2.9 A statically mounted Pratt & Whitney J58 engine with full afterburner on to dispose of the last of the SR-71 fuel.

Courtesy of NASA.

As a result, a thrust increase of more than 50% is common with afterburner use. A variable-area nozzle is necessary because of the differences in the volume of airflow, depending on whether or not the afterburner is operating, as exhaust gas is expanded to atmospheric pressure. In fact, this is a second combustion chamber positioned after the turbine and preceding the nozzle, which increases the temperature of the gas before the nozzle. The increase in temperature results in an approximately 40% increase in thrust at takeoff as well as a much larger increase post-takeoff.

The turbojet engine is also a reaction engine, in which expanding gases push hard against the front of the engine after having been sucked in and compressed. They then flow through the turbine, making it spin, before they shoot out of the rear of the exhaust and thus push the plane forward, similar to other jet engines, but in a far more powerful manner.

Figure 2.10 Schemata of a turbojet engine and the thrust equation terms.

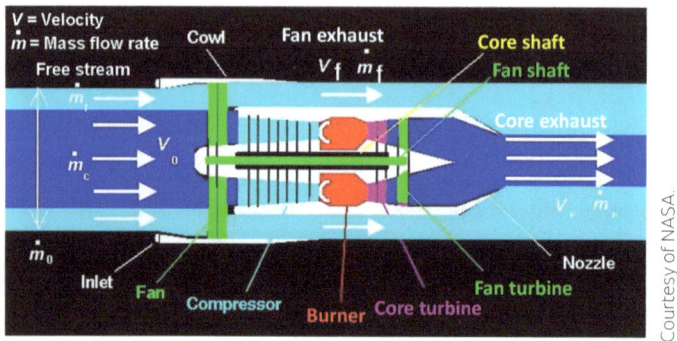

As you see in **Figure 2.10**, the fuel is injected twice into the engine. To calculate the thrust, you should consider whether the free stream conditions are denoted by a "0" subscript and/or the exit conditions by an "e" subscript. Therefore thrust, F, is equal to the mass flow rate m times the velocity V at the exit minus the free stream mass flow rate times the velocity.

$$\text{Thrust} = F = (m \times V)_e - (m \times V)_0$$

This equation is the most important and basic formula for propulsion calculations. For nerds: aerodynamicists often refer to the first term, $(m \times V)_e$, as the gross thrust and the second term, $(m \times V)_0$, as the ram drag, which is usually associated with conditions in the inlet.

Turboprops

Figure 2.11 A Turboprop engine at Kanoya Naval Air Base Museum.

When the jet engine is attached to a propeller, it is called a turboprop engine (**Figure 2.11**). The hot gases turn the turbine at the back, which then turns a shaft driving the propeller. This system is used in small airliners and transport aircraft but is usually considered to be more efficient than a piston engine, and more importantly, it works at higher altitudes than a piston engine.

Turboprop engines also use a compressor, combustion chamber, and turbine. Air under pressure is used to run the turbine, which then drives the compressor. In comparison with a turbojet engine, though, its propulsion is more efficient below speeds of about 800 km/h. Nowadays with modification, they can be efficient at higher flight speeds too. For this, they use propellers of a smaller diameter, but with a larger number of scimitar-shaped blades, with swept-back leading edges at the blade tips. Engines with these sorts of propellers are called propfans, or ultra-high-bypass turbofans. As can be seen in **Figure 2.12**, this is a type of aircraft engine related in concept to both the turboprop and turbofan, but distinct from both.

Figure 2.12 NASA/GE unducted fan or propfan.

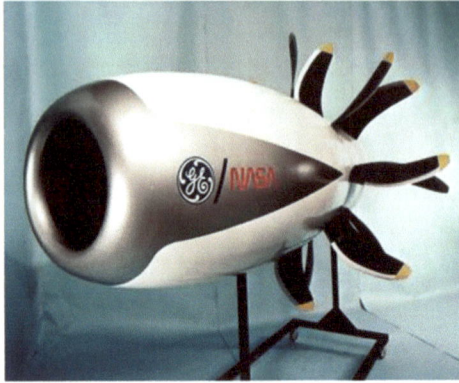

The engines are quite efficient; therefore, with increasing prices of jet fuel and the emphasis on reducing emissions, there is renewed interest in the propfan concept for jetliners that might come into service. Multiple design variations of this propfan were tested in conjunction with NASA in this decade.

Turboshafts

A turboshaft (see **Figure 2.13**) is another form of a gas-turbine engine. It operates like a turboprop system but powers a *helicopter* rotor instead. The speed of the helicopter rotor is designed to be independent of the rotating speed of the gas generator so the rotor speed can be kept constant even if the generator's speed varies for modulating the amount of power produced.

Figure 2.13 Schemata of a turboshaft engine.

Robysot/Shuttertock.com

Ramjets

Although a ramjet is the simplest and fastest jet engine and has no moving parts, you will rarely find one. This is because the application of ramjet is restricted by the fact that its compression ratio depends wholly on forward speed, which is a limitation. A ramjet starts working when the speed of the inlet air is above Mach 1 (more than 1224 km/h).

At those speeds, the jet forces, i.e., rams, air into its engine, as it is. In essence, it is a turbojet without any rotating machinery. As it does not develop static thrust or develops very little thrust below the speed of sound, it requires an assisted takeoff (i.e., the aid of another aircraft). Its primary use has been for guided missile systems and space vehicles.

At high flight velocities, the air hitting the engine intake produces enough compression for engine operation without an active compressor being required. This enables the ramjet to operate without any moving parts. As shown in **Figure 2.14**, a turbine is not necessary because there is no compressor.

Figure 2.14 Schematic of a ramjet.

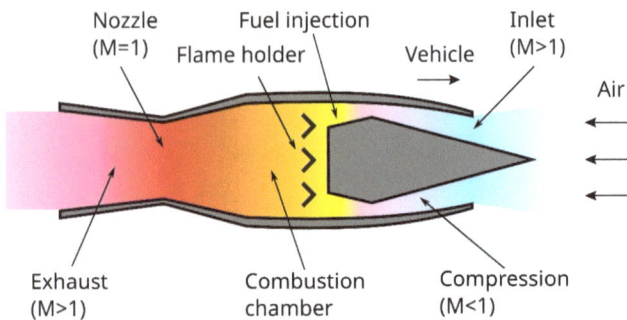

These engines have the advantage of being very simple. The obvious drawback is that they must have sufficient flight velocity before they can produce thrust. For this reason, they are used mainly on missiles augmented with rocket engines. As far as efficiency is concerned, ramjets begin to surpass afterburning turbojets at flight speeds approximating Mach 3.

Latest Jet Engine Developments

Jet engine development continues, and it is a very inspiring field of research. One of the ongoing challenges in this industry is producing 3D-printed engines. In 2018, GE announced "The world's largest jet engine took its maiden flight over the Mojave Desert on Tuesday, March 14. The GE9X engine has been known in the 3D printing world in particular as a case study for its 3D printed fuel nozzles. The GE9X also includes parts made from light and heat resistant ceramic composites and other new technologies. As a result the engine will be also up to 10 percent more fuel-efficient than the GE90, which is used for the current version of the 777 jet." (3D Printing Media Network, 2018.)

They will also use parts produced by 3D printing, which can produce complex shapes that would be difficult or impossible to make with conventional manufacturing techniques. 3D printing could eventually be used for making more engine parts. Similarly, maintenance of engines is important and costly as there are different development programs to predict potential failures of the engine and expect and repair them before occurrence to prevent catastrophes. In general, Etihad Engineering has made the best operational use cases for predictive maintenance on the overall aircraft architecture.

Conclusion

This chapter has given us a brief introduction to one of the key enablers of modern aviation, jet propulsion. In the next chapter, we will move from the sky to the stars.

Chapter 03

Spacecraft Propulsion

Figure 3.1 The April 12, 1981 launch at Pad 39A of STS-1, just seconds past 7 am, carried astronauts John Young and Robert Crippen into an Earth orbital mission scheduled to last for 54 h and ended with an unpowered landing at Edwards Air Force Base in California.

Courtesy of NASA.

Spacecraft propulsion refers to the methods used to propel spacecraft and artificial satellites. There are many different methods of spacecraft propulsion, each one with its own advantages and disadvantages. Not surprisingly, spacecraft propulsion continues to be actively researched. However, most spacecraft today are propelled by forcing a gas through a nozzle at the back of a rocket engine at very high speeds. All current spacecraft use chemical propulsion from either liquid (bipropellant) or solid fuels.

In Chapter 2, you learned that propulsion is a means of creating force that leads to movement. The systems designed for creating that force include a mechanical power supply (from an engine, motor, or muscles), and something for generating a force from it (such as a wheel and axles, propellers, a propulsive nozzle, wings, fins, or legs). Different types of air-breathing engines such as the Turbo Jet were discussed, but what happens when there is no air to breathe or power the engine? That is why different propulsion systems are used in space or somewhere with no significant amount of preexisting air.

History of the Rocket Engine

When it was lit, it made a noise that resembled thunder and travelled 100 Li (about 24 km). The place where it fell was burned and the fire extended more than 600 meters. These iron nozzles, the flying powder halberds that were hurled, were what the Mongols feared most.

These words were written by Father Antoine Gaubil in 1739 in conjunction with his book on Genghis Khan. They describe how a Chinese town in 1232 successfully defended itself against 30,000 invading Mongols by means of rocket-propelled fire arrows. They are an example of historical evidence showing that rocketry was born and developed in the Orient many centuries ago. It is reasonably clear that the Chinese manufactured black powder at least as early as 600 A.D. and subsequently used this mixture of charcoal, sulfur, and saltpeter as a rocket propellant.

Over the centuries, rockets slowly spread to the West for use as military weapons. They were much improved as a barrage missile by Sir William Congreve in England in the early 1800s. The "rocket's red glare" observed by Francis Scott Key in 1812 at Fort McHenry was produced by a Congreve rocket. However, it was not until the end of the nineteenth and the beginning of the twentieth centuries that rockets were understood from a technical point of view and their true engineering development began.

The Soviet Union was the first into space with an artificial satellite, Sputnik I, on October 4, 1957, and then with a human in orbit, Yuri Gagarin, on April 12, 1961. Thus, from a historical perspective, it is fitting that the first true rocket scientist was a Russian, Konstantin Eduardovitch Tsiolkovsky, born in September 1857 in the town of Izhevsk Oye. As a young student, he absorbed physics and mathematics and was tantalized by the idea of interplanetary space travel. In 1876, he became a schoolteacher in Borovsk and in 1882 moved to the village of Kaluga. There in virtual obscurity, he worked on theories of space flight and in March 1883 hit upon the idea of reactive propulsion.

Working without any institutional support, Tsiolkovsky gradually solved some of the theoretical problems of rocket engines. **Figure 3.2** shows his design of a rocket, fueled with liquid hydrogen (H_2) and liquid oxygen (O_2), which was published in the Russian magazine *Science Survey* in 1903 (the same year as the Wright brothers' successful first powered airplane flight). The fact that Tsiolkovsky knew to use the high-specific-impulse combination of H_2-O_2 testifies to the sophistication of his rocket theory.

Tsiolkovsky was neither an experimentalist, as he did not have the money to develop a laboratory, nor an engineer. So, he did not conduct any practical experiments or generate any design data. Nevertheless, he was the first true rocket scientist, and he worked incessantly on his theories until his death on September 19, 1935. In his later life, his contributions were finally recognized, and in 1919, he became a member of the Socialist Academy (forerunner of the U.S.S.R. Academy of Science) and was subsequently granted a government pension.

Figure 3.2 Tsiolkovsky's drawing of the evolution of a rocket shows his systematic thinking.

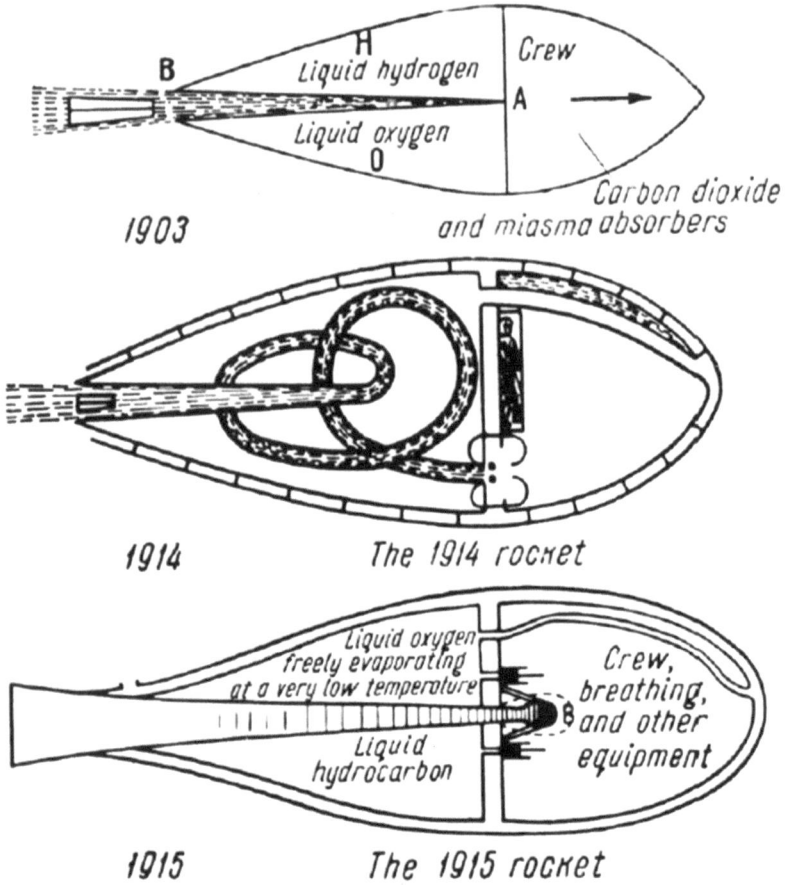

B

H
Liquid hydrogen

Crew
A →

Liquid oxygen
O

Carbon dioxide
and miasma absorbers

1903

The 1914 rocket

1914

Liquid oxygen
freely evaporating
at a very low temperature

Crew,
breathing,
and other
equipment

Liquid
hydrocarbon

1915   The 1915 rocket

NASA/Marshall Space Flight Center.

At the turn of the century, rocketry arrived in the US in the form of Dr. Robert H. Goddard, who was born in Worcester, Massachusetts, on October 5, 1882. His life had many parallels to Tsiolkovsky's: he too was an avid physicist and mathematician; he too was convinced that rockets were the key to space flight; and he too worked in virtual obscurity for most of his life. But there was one difference. Whereas Tsiolkovsky's contributions were purely theoretical, Goddard successfully melded theory into practice and developed the world's first working liquid-fueled rocket.

Goddard was educated thoroughly and subsequently became a professor of physics locally at Clark University, where he began to seriously apply science and engineering in the service of his childhood dreams of space flight. He too worked out that liquid H_2 and O_2 would be very efficient rocket propellants, and he pursued these ideas during a leave of absence at Princeton University during 1912–1913. In July 1914, Goddard was granted patents on rocket combustion chambers, nozzles, propellant feed systems, and multistage rockets.

In 1917, Goddard obtained a small grant ($5000) from the Smithsonian Institution in Washington, permanently establishing him in a rocketry career. This grant led to one of the most historic documents of rocket engine history—a monograph entitled *A Method of Reaching Extreme Altitudes*, which was published as part of the Smithsonian Miscellaneous Collections in 1919.

This book was a scholarly and authoritative exposition of rocket principles, but at that time, few people seized upon his ideas. Goddard increased his laboratory activities at Worcester in the early 1920s. Here, after many tests and much engineering development, he successfully launched the world's first liquid-fuel rocket on March 16, 1926. A picture of Goddard standing beside this rocket is shown in **Figure 3.3**. The vehicle was 3 m long with a motor at the very top and was fed liquid oxygen and gasoline through two long tubes that led from the propellant tanks toward the vehicle's rear (below Goddard's outstretched arm in the figure). The conical nose on the fuel tanks was simply a deflector to protect the tanks from the rocket nozzle exhaust. The rocket reached a maximum speed of 97 km/h and flew 56 m.

Although modest in performance, this flight contributed to rocketry just as what the Wright brothers' December 17, 1903 flight contributed to aviation. Ultimately, it brought Goddard to the attention of Charles A. Lindbergh, when he had considerable stature due to his 1927 trans-Atlantic flight. Lindbergh was subsequently able to convince the Daniel Guggenheim Fund for the Promotion of Aeronautics to give Goddard a $50,000 grant to further pursue rocket engine development.

Figure 3.3 Rocketry pioneer Robert H. Goddard and his first liquid-fueled rocket.

© ESA.

Suddenly, Goddard's operation magnified, and in 1930 he and his wife moved to a more suitable testing location near Roswell, New Mexico. There, for the next 11 years, Goddard made bigger and better rockets although still in an atmosphere of obscurity. The government was simply not interested in any form of jet propulsion research during the 1930s. Additionally, Goddard, like the Wright brothers, imposed a blanket of secrecy on his data for fear of others pirating his designs. Until July 1945, this group developed jet-assisted takeoff units for seaplanes and worked on a variable-thrust rocket engine.

On August 10, 1945, Dr. Robert H. Goddard died in Baltimore. Recognition for his contributions to the development of modern rocketry and realization of their importance came late. Indeed, only in the political heat of the post-Sputnik years did the US really pay homage to him. In 1959, he was honored by Congress, and in the same year, he received the first Louis W. Hill Space Transportation Award of the Institute of Aeronautical Sciences (now the American Institute of Aeronautics and Astronautics). On May 1, 1959, the new NASA Goddard Space Flight Center in Greenbelt Maryland, was named in his honor. Finally, in 1960, the Guggenheim Foundation and Mrs. Goddard were given $1,000,000 by the government for use of hundreds of Goddard's patents.

During the 1930s, and completely independent of Goddard's operations, another small group in the US was developing rockets. This was the American Rocket Society (ARS) that was originally founded in March 1930. This society not only published technical papers but also built and tested actual vehicles. Its first rocket, burning liquid oxygen and gasoline, was launched on May 14, 1933, in Staten Island New York, and reached 76 m. Following this and up to World War II, the ARS was a public focal point for small rocket research and development, but without any government support.

After the beginning of World War II, much ARS experimental activity was splintered and absorbed by other activities around the country. Then, the ARS and the Institute of Aeronautical (by that time Aerospace) Sciences were merged to form the present American Institute of Aeronautics and Astronautics. As a brief example of how the threads of the history of flight are interwoven, in 1941, members of the ARS formed a company, Reaction Motors, Inc., which went on to design and build the XLR-11 rocket engine. This was the engine that powered the Bell X-I and pilot Chuck Yeager to the first manned supersonic flight on October 14, 1947. You can see this engine on display at the National Air and Space Museum (**Figure 3.4**).

Figure 3.4 V-2 Rocket leaving the firing table.

Crown Copyright/https://en.m.wikipedia.org/wiki/File:V-2_lift-off.jpg

The early history of rocket engines forms a geographical triangle with one vertex in Russia (Tsiolkovsky), another in the US (Goddard), and the third in Germany represented by Hermann Oberth, who was born in Transylvania on July 25, 1894, and later became a German citizen.

In 1923, Oberth published his own work on the theory of rocket engines entitled *The Rocket into Planetary Space*. This was a rigorous technical text that laid the basis for the development of rockets in Germany. In order to foster Oberth's ideas, the German Society for Space Travel was formed in 1927 and began experimental work in 1929. (This society provided the template for the later American Rocket Society.)

Oberth's ideas had a catalytic effect, especially on some of his students such as Wernher Von Braun and resulted in an almost explosive development of rocketry in Germany in the 1930s. This work, with Von Braun as the technical director, culminated in the development of the German

V-2 rocket of World War II fame (see **Figure 3.4**). Although an instrument of war, the V-2 was the first practical long-range rocket in history. Powered by liquid oxygen and alcohol, it was 14 m long and 1.65 m in diameter and weighed 12,500 kg. It was the first vehicle made by humans to fly in space with altitudes above 80 km and a range of 320 km. The missile reached supersonic speeds during its flight within the atmosphere.

Toward the end of World War II, Russian and US forces captured hundreds of V-2s and shipped them back to their respective countries. As a result, the ancestry of all modern rockets derives from the V-2 and back beyond that to Von Braun and eventually to Hermann Oberth. The full development of modern rockets that culminated in a huge Saturn booster for the Apollo program is a huge story that goes beyond the scope of this book. However, we will outline its early history so you can appreciate the technical aspects of rocket engines discussed later.

During and after World War II, many rocket-powered aircraft were built so high speed flight could be explored. These included the X-1A, which broke the sound barrier, and the X-15, which was a rocket-powered airplane (**Figure 3.5**).

The X-15 was a joint program by NASA, the Air Force, and the Navy. It is sometimes called the most remarkable of all the rocket research aircraft operated ever.

This rocket is composed of an internal structure of titanium and a skin surface of a chrome-nickel alloy known as Inconel X. The X-15's first powered flight took place on September 17, 1959. Because of the large fuel consumption of its rocket engine, the X-15 was air launched from a B-52 aircraft at about 45,000 ft and speeds upward of 500 mph. The airplane first set speed records in the Mach 4-6 and also set an altitude record of 354,200 feet (67 miles) and provided an enormous wealth of data on hypersonic air flow, aerodynamic heating, control and stability at hypersonic speeds. The program's final flight was performed on October 24, 1968.

—Steve Garber, NASA History.

Figure 3.5 X-15 Hypersonic Aircraft/Rocket.

Courtesy of NASA

At the forefront of modern aerospace innovation is the Boom Overture, a supersonic aircraft designed to redefine commercial air travel. Inspired by the speed and engineering achievements of the X-15, the Boom Overture is planned to reach Mach 1.7, carrying up to 80 passengers over distances of 4890 mi. By combining advanced materials and aerodynamic designs, this next-generation aircraft aims to make supersonic travel a viable and sustainable option for passenger aviation.

Projects like the Boom Overture not only revisit the glory of past hypersonic programs but also push the boundaries of what is possible in the future of aerospace engineering. This blend of historical and modern advancements demonstrates how aerospace technology evolves, from high-speed research aircraft to innovative solutions for commercial and interstellar applications.

Rocket engines mix the fuel and an oxygen source called an oxidizer and then explode them in a combustion chamber to produce a working fluid of hot exhaust. This working fluid is then passed through a nozzle in order to accelerate its flow and produce thrust. This working fluid is essentially different from the one used in a turbine engine or a propeller-powered aircraft discussed in Chapter 2, which uses air from the atmosphere as the working fluid, but rockets use the exhaust gases from combustion. As there is no atmosphere in outer space, turbines and propellers cannot work there. It is interesting, isn't it? Rockets work in space, but turbine engines and propellers do not!

The most common rockets are **liquid rockets** (**bipropellant**) and solid rockets. For liquid rockets, the propellants (fuel plus oxidizer) are stored separately in liquid form. For solid rockets, the propellants are mixed together in powder form and packed into a solid cylinder.

At normal temperatures, the propellants do not burn. However, they do burn when they are ignited in the combustion chamber. Once started, the burning continues until all the propellant is exhausted. In a liquid rocket, the thrust can be stopped and/or controlled by turning off the flow of propellants. In a simple solid rocket, it is not possible as all the propellant is already in the combustion chamber. However, nowadays, ways have been found to extinguish solid propellant.

The electric engine (ion engine) is a new type of rocket. Its working fluid comprises many very small, charged particles called ions. This working fluid is accelerated by electrostatic forces instead of combustion. Even though the thrust they produce is very small, ion engines can do that for a very long time because the mass flow rates are very small. Compared to chemical rockets, ion engines have a very high specific impulse (an efficiency measurement showing how effectively an engine uses its fuel).

The nuclear thermal engine is another new type of rocket engine, in which the working fluid (any gas) is accelerated by a continuous source of heat from a nuclear reactor. The heating is provided when the working fluid is passed over or through the reactor and from there exits directly through the nozzle. Because the temperature of the exhaust can be much higher than that of a typical chemical rocket, nuclear thermal engines produce a much-increased exit velocity. They are also predicted to have very high specific impulse because they only use a single working fluid. Nuclear thermal engine development occurs under the umbrella of Project Prometheus.

The working fluid in the simplest rocket engine is air-pressurized by a pump so it will accelerate—in a stomp rocket (toy balloon). Since the weight flow of air (i.e., weight of the air that has flowed) is tiny, this type of rocket engine does not produce much thrust. A bottle rocket (water rocket), which many of you may know, uses water as the working fluid, which is pressurized by air to accelerate it. Bottle rockets generate more thrust than stomp rockets because water is much heavier than air. If you do not know about these cool engine types, you might find it useful to research and build them so you can understand them better.

Forces in Rockets

The rocket's propulsion system generates thrust again according to the dictates of Newton's third law of motion: for every action, there is an equal and opposite reaction (see **Figure 3.6** for the force diagram of a rocket). An engine works on the gas or liquid working fluid and accelerates it through the propulsion system. Reaction to that acceleration produces the thrust force on the engine. The working fluid is expelled from the engine in one direction so that the thrust force moves the engine in the opposite direction.

Forces are vector quantities, which means that they are characterized by both size (magnitude) and direction. This means that both the size and the direction of a force must be taken into account.

Usually, thrust is directed along the rocket's longitudinal axis through its center of gravity. Sometimes, though, the exhaust nozzle and the thrust direction can be rotated (gimbaled), so the rocket can be maneuvered by using the torque (rotational force) about the rocket's center of gravity. The thrust's size can be calculated with the General Thrust Equation:

Figure 3.6 Force diagram of a rocket.

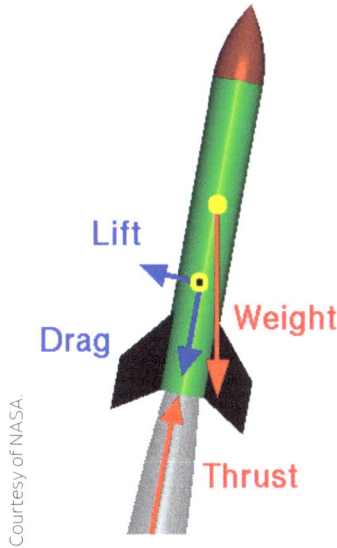

The size of the thrust depends on the mass flow rate of the working fluid through the engine and the velocity and pressure of the existing working fluid.

The propulsion system's efficiency is characterized by the specific impulse where

The specific impulse is the ratio of the thrust produced to the mass flow of the propellants.

Rocket Systems

Studying rocket systems is an excellent way for you to learn the basics about forces and how an object responds to external forces. Rockets in flight are subjected to many forces: weight, thrust and aerodynamics.

In **Figure 3.7**, the rocket's outer skin has been removed to reveal its constituent parts, allowing those parts to be designed and analyzed effectively. The ones with the same function are grouped into systems, the major ones being for structure, payload, guidance, and propulsion.

Figure 3.7 Subsystems of a rocket.

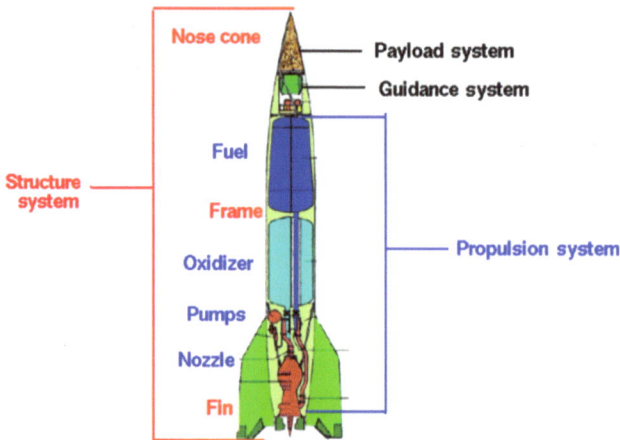

The **structural system** (frame), like the fuselage of an airplane, is made from very strong but lightweight material, like titanium or aluminum. It is attached to long stringers running between the top and bottom and to hoops running around the circumference. The skin's thermal protection keeps out the heat generated by air friction during flight and keeps in the cold temperatures required for some fuels and oxidizers. Fins are also attached to some rockets at their frame bottom to provide controllability during flight.

The nature of a rocket's **payload system** depends on the rocket's mission. Payloads vary from the earliest celebratory fireworks for holidays to several hundreds of kilograms of explosives for the German V2. After World War II, guided ballistic missiles had nuclear warheads for payloads. Rockets were modified to launch satellites with wide-ranging missions covering communications, weather monitoring, spying, planetary exploration and observatories such as the Hubble Space Telescope. Some rockets were even developed for special human payloads: to launch people into orbit around Earth and onto the surface of the Moon.

The **guidance system** usually includes highly sophisticated sensors, onboard computers, radars, and communication equipment for maneuvering the rocket in flight and is explained in the next section.

Guidance for Rockets

Guidance in rockets is accomplished by a very sophisticated array of sensors, onboard computers, radar, and communication equipment that modern rockets carry. The guidance system's purpose is to control the rocket during the flight based on its current speed and direction of travel and to determine what maneuvers it needs to maintain a desired trajectory. A range of different mechanisms have been developed for this.

The movement of something in flight can be described as a combination of the change in position of its center of gravity together with its rotation about its center of gravity (i.e., its center of mass like a gyroscope). Each of the control mechanisms produces a torque about the rocket's center of gravity which then makes the rocket rotate in flight. By understanding the forces acting on the rocket and their resultant motion, the rocket guidance system can be programmed so that the rocket is able to fly into the target orbit.

Movable fins at the rear of the rocket were used on early rockets and are currently used on air-to-air missiles. These fins help adjust the aerodynamic forces on the rocket that act through the center of pressure, which is not normally located in the same place as the center of gravity. Actually, it is this difference in location that generates the torque about the center of gravity. **Figure 3.8** shows what happens if the trailing edge of a movable fin is deflected to the right; for instance, the resulting aerodynamic force will move the nose of the rocket to the right.

Figure 3.8 Rocket maneuvering systems.

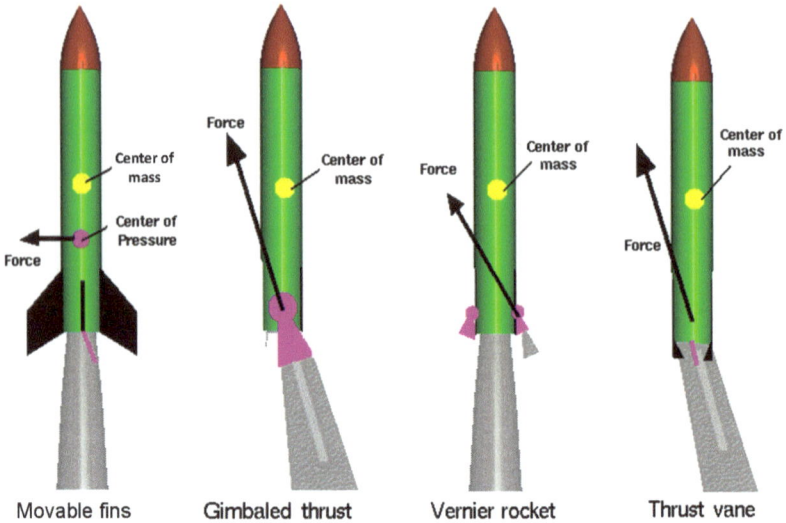

Movable fins   Gimbaled thrust   Vernier rocket   Thrust vane

Courtesy of NASA.

Most of the time, modern rockets control their torque by rotating their nozzle in order to produce controlled torque. Sometimes, the system uses a gimbaled thrust instead so the rocket's exhaust can be swiveled from side to side, and in the process, the direction of the thrust is changed relative to the rocket's center of gravity. **Figure 3.8** shows what happens if the rocket nozzle is pivoted to the right; for example, the resultant thrust force will move the rocket's nose to the right.

A different method was used in some older rockets. The Atlas missile for instance used additional small rocket engines (called Vernier rockets) at the bottom of the main rocket to generate the control torque when required. **Figure 3.8** shows what happens if a Vernier rocket engine on the right is fired. It will move the nose of the larger rocket to the right. Vernier rockets are not used much anymore as they require extra fuel and plumbing weight.

In other early rockets like the V2 and Redstone Rocket, the thrust was deflected by small thrust vanes positioned in the rocket's exhaust stream to produce a controlling torque. This is a simple guidance system that uses a Vane engine together with a PIGA. PIGA is basically making use of a gyroscope to control the fuel and angle. It is valued for its high sensitivity and accuracy in conjunction with operation over a wide acceleration range. PIGA is still considered the premier instrument for strategic-grade missiles.

Many different methods have been developed for this—some of which are illustrated in **Figure 3.9** and discussed so far in this section. The guidance system also provides some level of controllability so the rocket does not tumble in flight. In fact, the biggest proportion of a full-scale rocket is taken up by its propulsion system, which you will learn more about later in this chapter.

Figure 3.9 Pendulous integrating gyroscopic accelerometer.

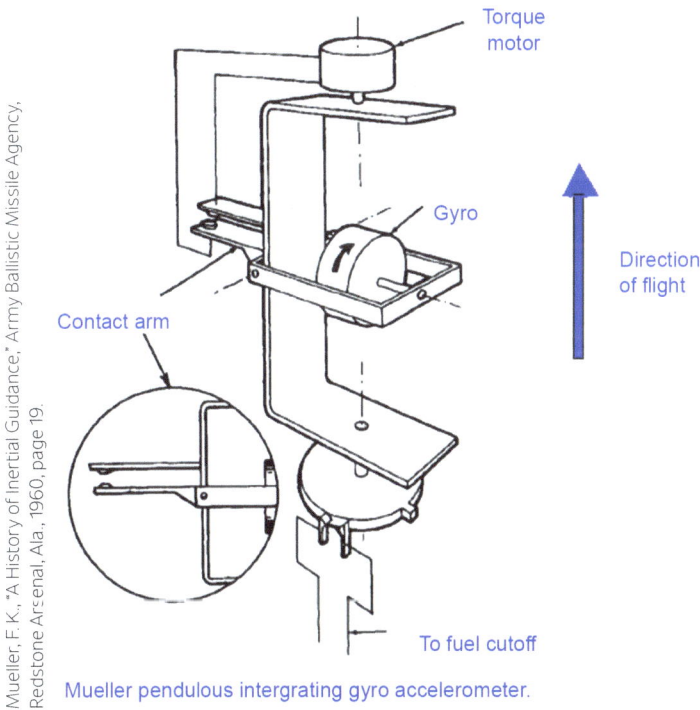

Mueller, F. K., "A History of Inertial Guidance," Army Ballistic Missile Agency, Redstone Arsenal, Ala., 1960, page 19.

Mueller pendulous intergrating gyro accelerometer.

Different Types of Rocket Engines: Bipropellant Engine

Well-known applications of liquid rocket engines (bipropellant engines) (**Figure 3.10**) can be found in the Space Shuttle (for putting humans into orbit), in many unmanned missiles (for putting satellites into orbit), and in several high-speed aircraft used for research purposes after World War II. The thrust produced by a liquid rocket engine depends on the mass flow rate of the propellant through the engine, the exit velocity of the exhaust, and its pressure at the nozzle exit. But all these variables depend on the nozzle design. The nozzle's narrowest area is referred to as its throat, which functions to choke the hot exhaust flow from the combustion chamber to the rocket's exit.

Figure 3.10 A schematic of a liquid rocket engine and the terms of the rocket equation.

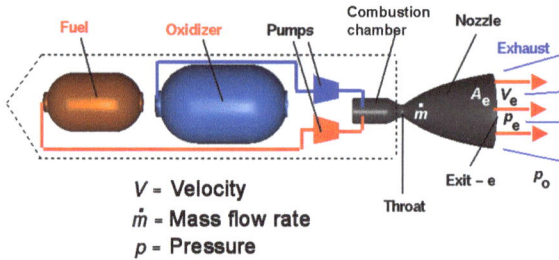

V = Velocity
$\dot{m}$ = Mass flow rate
p = Pressure

Thrust = $F = \dot{m} V_e + (p_e - p_0) A_e$

The Mach number in the throat is equal to 1.0, and the mass flow rate m dot ($\dot{m}$) is governed by the area of the throat. The exit velocity V_e and the exit pressure P_e are governed by the ratio of the throat's area to the exit area A_e. Because the exit pressure is only equal to the free stream pressure in some design conditions, we must use the longer version of the Generalized Thrust Equation to describe the system's thrust, F. If the free stream pressure is given by P_0, the thrust equation then becomes:

$$F = \dot{m} \times V_e + \left(P_e - P_0\right) \times A_e$$

Analysis of rocket performance efficiency is greatly simplified because of the efficiency parameter (specific impulse), which works for both types of rockets. How to mix and burn the fuel and oxidizer without blowing out the flame is very complex—a rocket scientist is needed to figure that out!

Solid Rocket Engine

Well-known applications of solid rocket engines (**Figure 3.11**) are found in air-to-air and air-to-ground missiles, in model rockets, and as boosters for satellite launchers. Their solid propellant is packed into a solid cylinder with a hole that serves as the combustion chamber. Once ignited, combustion takes place on the surface of the propellant and creates a flame-front that burns into the mixture. The area of the flame front governs the amount of exhaust gas produced so engine designers vary the hole shapes in order to control the thrust for particular engines. As for liquid rocket engines, the hot exhaust gas passes through a nozzle to accelerate its flow and produce thrust according to Newton's third law of motion which will be explained more in Chapter 5.

Figure 3.11 A schematic of a solid rocket engine showing terms of the rocket equation.

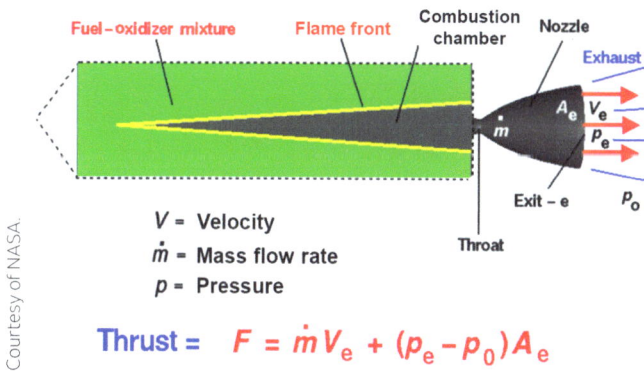

Courtesy of NASA.

V = Velocity
$\dot{m}$ = Mass flow rate
p = Pressure

Thrust = $F = \dot{m} V_e + (p_e - p_0) A_e$

The amount of thrust produced by this type of rocket also depends on the nozzle's design. The physics of the Mach number in the throat, the mass flow rate m dot $(\dot{m})$, A_e, V_e and P_e is the same as it is for liquid rocket engines with the result that the thrust equation for solid rockets is also:

$$F = \dot{m} \times V_e + (P_e - P_0) \times A_e$$

The convenient upshot is that the thrust equation works for both solid and liquid rocket engines.

Electric Propulsion

Electric propulsion applies to all the different ways that a propellant can be accelerated by using electrical power. It is most easily achieved by using electrical heating instead of the chemically produced heat from a combustion engine. Placing your pressure cooker on an electric cooker hotplate is an example of this. Electrical heating can be achieved by using a controlled electrical discharge (arc) to heat a stream of gas. It can also be achieved by using other more sophisticated and more efficient ways to obtain a fast jet of gas. A diagram of how a gridded electrostatic ion engine (i.e., DS-1 is an ion engine built by NASA) works is shown in **Figure 3.12**.

The DS-1 ion propulsion system uses a hollow cathode to produce electrons to collisionally ionize xenon. The Xe+ is electrostatically accelerated through a potential of up to 1280 V and emitted from the 30-cm thruster through a molybdenum grid. A separate electron beam is emitted to produce a neutral plasma beam.

Figure 3.12 DS-1—How the ion engine works, by NASA GRC.

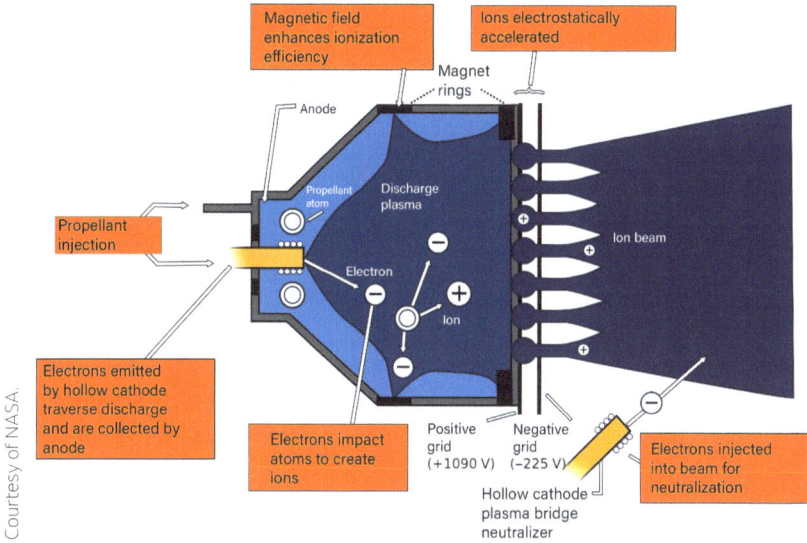

Magnetic field enhances ionization efficiency

Ions electrostatically accelerated

Magnet rings

Anode

Propellant atom

Discharge plasma

Propellant injection

Electron

Ion beam

Electrons emitted by hollow cathode traverse discharge and are collected by anode

Electrons impact atoms to create ions

Ion

Positive grid (+1090 V)

Negative grid (−225 V)

Hollow cathode plasma bridge neutralizer

Electrons injected into beam for neutralization

Courtesy of NASA

To understand how this works, one of the classic home science experiments is to lift a small piece of paper with a plastic rod (usually a biro), after rubbing the biro on some woolen material to make it electrically charged. That creates a static electric field, and then the biro attracts the paper. If you try this and it does not work, it is because you did not charge the rod enough or did not rub the biro enough. Another reason may be because the biro's plastic is too conductive or loses the charge too quickly.

Now imagine that instead of a plastic rod, you have a grid with relatively large holes and instead of a piece of paper you have a lot of tiny balls. If the grid is charged statically, the balls will float toward it. The force attracting them is proportional to the charge on the grid. If the charge is increased, the balls will move toward the grid at a higher speed. Once the speed is high enough, most balls will not stick to the grid but pass straight through it and then keep on going.

This is in effect what happens in electrostatic propulsion: tiny balls of ionized particles of propellant are launched past the grid, thereby pushing the satellite or space vehicle in the opposite direction. In another form of electric propulsion, an electric field for setting the particles in motion plus a magnetic field for accelerating them are used in combination. When an electric field alone is used, charged particles of the polarity opposite to the grid's polarity provide propulsion.

However, when electric and magnetic fields are combined, both polarities are accelerated. In a magneto-electric thruster, the electric field makes the charged particles move in opposite directions. Using what is known by physicists as the Lorentz force, a magnetic field affects a moving charge by pulling the charged and moving particle sideways. These different effects of electric and magnetic fields are used in combination to accelerate and eject the charged particles from the thrusters, to push the satellite in the opposite direction.

The various types of thrusters are characterized by different performance levels. Some are better for missions requiring greater thrust levels to reduce the overall trip time. Others are better for high-precision positioning. Yet others are better when minimal propellant is required.

Electric propulsion has been studied in parallel with chemical propulsion for many years. It has always fascinated science-fiction writers who have used it to power their spaceships, the most famous one being the Discovery in the film *2001: A Space Odyssey*. Until recently, its application in real life has been held back by the low availability of electrical power onboard spacecraft. Nowadays, electric-propulsion thrusters have become more efficient, and together with the much greater electrical power available on today's spacecraft, this has made electric propulsion a real possibility.

The Past

The NASA Glenn Research Center has led the study of electric propulsion since it began studying ion propulsion in the 1950s. On July 20, 1964, it first operationally tested an ion propulsion system (IPS) in space with the Space Electric Rocket Test 1 (SERT 1), which returned to Earth after successfully completing its goal of a 31-min operation. This successful ion-propulsion test was followed by many others.

The Present

The core technologies enabling major enhancements in IPSs are primarily developed at NASA's Energetics Program at Glenn. Many areas of research are involved: strong development and test efforts, computer modeling, hollow cathode technology advances, and enhanced ion optics technologies. The study of plasma's basic principles allows the creation of models that will reduce the cost of designing ion thrusters in the future. Components that can last longer or operate in extreme conditions can now be produced because of the further development of component technologies.

A high-power nuclear electric-powered ion propulsion system for the Jupiter Icy Moons Orbiter (JIMO spacecraft) has been developed. In **Figure 3.13**, a blue beam of solar electric propulsion (SEP) produces 70 kps of thrust. This high-power electric propulsion ion thruster is unique and is currently the most powerful inert gas ion thruster that has ever been built. With power levels of 40 kW, exhaust velocities in excess of 90,000 mps (over 200,000 mph) can be produced.

Figure 3.13 Blue beam of Xenon ions ejected from a SEP, an ion engine at approximately 70 kps.

Courtesy of NASA.

The Future

To extend the operational life of satellites and reduce launch and operation costs, more and more companies are using satellites propelled electrically. The resultant savings can be passed on to consumers. The development of SEP technologies has the potential to improve space transport efficiencies enabling deep space travel.

Staged Launchers

Most of a rocket's weight comes from the weight of propellants. Once the propellants are burned off during powered ascent, most of the weight of the vehicle reduces to that of the near-empty tanks plus the structure that was required for a fully loaded vehicle. To further lighten the vehicle prior to orbital velocity, a portion of it is discarded in a process called staging, which can be serial or parallel—or both.

For *serial staging*, a small, second-stage rocket is positioned on top of a larger first (i.e., upper) stage rocket. The functionality of this staging is demonstrated in **Figure 3.14**. At launch, the first stage is ignited and powers the vehicle on its ascent until its propellant is exhausted. At this point, the first-stage engine is extinguished and separates from the second stage, which is then ignited. The vehicle's payload is carried on top of the second stage into orbit. The Saturn V Moon Rocket used a three-part serial staging system to get into orbit around Earth. Its discarded stages were considered expendable and never recovered.

In *parallel staging*, as shown in **Figure 3.15**, a group of small first stages are strapped onto a sustainer rocket. At launch, all engines are ignited and the strap-on rockets are discarded when their propellants are extinguished. The vehicle's payload is carried on top of the still-burning sustainer rocket into orbit. The Space Shuttle used parallel staging to get into orbit around Earth. Its discarded solid rocket boosters were not considered expendable and were retrieved from the ocean for refilling with propellant and reuse on another Shuttle launch.

Figure 3.14 Serial staging for powering the vehicle ascent.

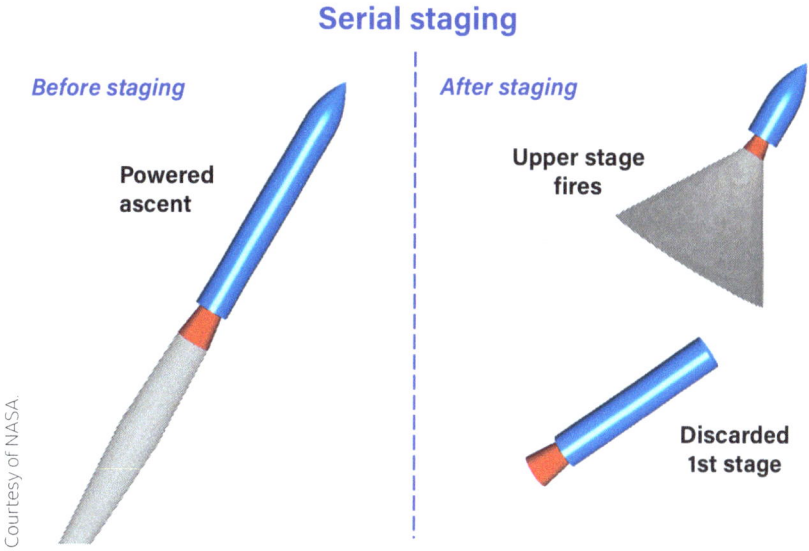

Serial staging

Before staging

Powered ascent

After staging

Upper stage fires

Discarded 1st stage

Courtesy of NASA.

Figure 3.15 Parallel staging for powering the vehicle ascent.

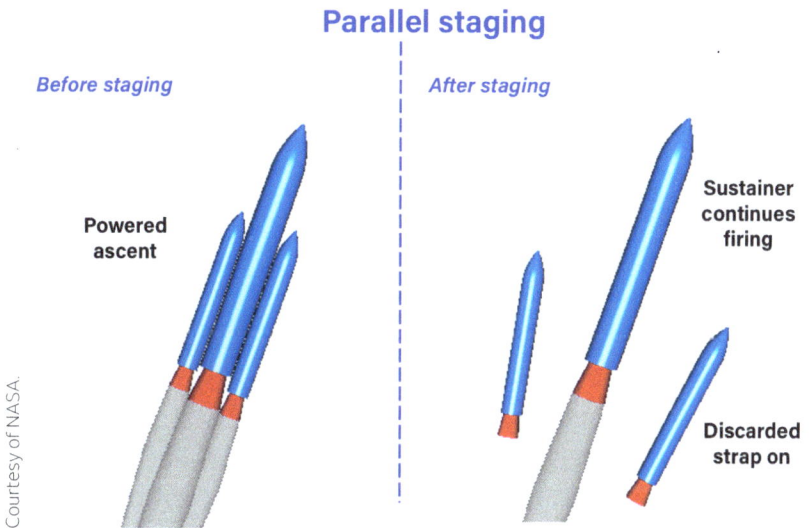

Parallel staging

Before staging

Powered ascent

After staging

Sustainer continues firing

Discarded strap on

Courtesy of NASA.

Some vehicles, like the Titan III and Delta II, use a combination of both serial and parallel staging. The Titan III uses a liquid-powered, two-stage Titan II for a sustainer rocket plus two solid rocket strap-ons for launching. After the solid strap-ons are discarded, the first stage of the sustainer engine of the Titan II is burned until its fuel is exhausted and then discarded, leaving the second stage of the Titan II to burn and carry its payload into orbit. The Delta II provides another example of a three-stage rocket.

Conclusion

This chapter has given a brief introduction to the history of rocket development, current major types of rocket architectures, and a brief preview into potential future propulsion developments. In the next few chapters, we will examine the wide array of space activities that have been enabled by rocket propulsion, including human space exploration.

Chapter 04

Space Exploration

Figure 4.1 Apollo 11 would not have been possible without the geopolitical rivalry that drove a race to the Moon.

Courtesy of NASA.

It's human nature to stretch, to go, to see, to understand. Exploration is not a choice, really; it's an imperative.

—Michael Collins, flew on Gemini 10 and Apollo 11

In 1961, Soviet cosmonaut Yuri Gagarin became the first human to visit space, and eight years later, US astronaut Neil Armstrong became the first man to walk on the Moon. What did it take for the human race to accomplish these feats, and what kind of scientific discoveries had to be made to send people into that great unknown and then bring them safely home again?

Of course, space exploration did not stop there. In the years since the space race, we have continued to learn more and more about the great void beyond our planet's atmosphere. Whether we are sending astronauts to do experiments aboard the International Space Station (ISS) or sending rovers to explore Mars, there is always something going on with the world's space programs.

In this chapter, you will learn about the present and future of space exploration. In order to further the human race, we need to continue to go further than we have ever gone before. Despite whatever setbacks we may face, space exploration is a necessity and the best way for us to see and experience our tomorrows.

A Brief History of Space Exploration

Humans have dreamed about spaceflight since antiquity. As you know, almost a thousand years ago rockets were being used in ceremonies and warfare by the Chinese, but it was not possible to build rockets that were powerful enough to overcome the force of gravity and reach orbital velocities to allow human exploration of space until the latter half of the twentieth century.

After the first rocket engine was designed by Tsiolkovski in the 1930s and 1940s, Nazi Germany began preparing for the use of long-distance rockets as weapons. Late in World War II, London was attacked by V-2 missiles which had a range of over 320 km and had arched across the English Channel at a height of 100 km at a speed of more than 5500 km/h.

Post-World War II, the US and the Soviet Union each set up their own missile programs. The race was on! The Soviets launched the first artificial satellite, Sputnik 1, into space on October 4, 1957, and then orbited Lt. Yuri Gagarin around Earth in Vostok 1 in April 1961—the first human to do so. This flight lasted 108 min and reached an altitude of 327 km (roughly 202 mi). The first US satellite, Explorer 1, flew into orbit on January 31, 1958, and Alan Shepard became the first American to fly into space in 1961. The first American to orbit Earth was John Glenn in February 1962.

President John F. Kennedy set a national goal for the US in 1961 of landing a man on the Moon and returning him safely to Earth within a decade. This goal was achieved on July 20, 1969, by NASA when astronaut Neil Armstrong set foot on the lunar surface during the Apollo 11 mission, thereby taking a giant leap for mankind. In total, six Apollo missions explored the Moon between 1969 and 1972.

Before astronauts ever landed on it throughout the 1960s, the Moon was probed and photographed by unmanned spacecraft, Earth-orbiting communications, and navigation satellites that were in everyday use. The surface of Mars was being mapped in the early 1970s by the Mars-orbiting Mariner spacecraft. By the end of the 1970s, detailed images of Jupiter and Saturn, their rings, and their moons had been sent to NASA from the Voyager spacecraft.

Human spaceflight highlights from the 1970s included Skylab, US' first space station (**Figure 4.2**) and the Apollo Soyuz Test Project, which was the world's first space mission to be internationally crewed—by Americans and Russians. It became possible to transmit television programs via communications satellites and for people to pick up those satellite signals on their home dish antennas in the 1980s.

The ozone hole over Antarctica was discovered by a satellite. Forest fires were pinpointed. We saw the nuclear powerplant disaster at Chernobyl in 1986. Astronomical satellites found new stars and gave us new views of the center of our galaxy. Reliance on the reusable shuttle for most civilian and military space missions began with the launch of the space shuttle Columbia in April 1981 and continued until the explosion of the shuttle Challenger during launch, which killed its crew of seven in January 1986. Twenty-four successful shuttle launches fulfilled many scientific and military requirements during those five years.

Figure 4.2 The International Space Station in orbit.

Courtesy of NASA

US' space program was re-evaluated after the explosion of the Challenger. The goal to re-evaluate was morphed into one of ensuring that each satellite had a suitable launch system available when it was scheduled to fly. Nowadays, this is achieved by the availability of more than one launch method and launch facility. Satellite systems are also designed for compatibility with multiple launch systems. The value of satellites in modern conflicts was demonstrated in the Gulf War, when Allied forces could achieve a decisive advantage by using their control of the "high ground" of space.

Satellites during the war provided information on enemy troop formations and movements. They also gave early warnings of enemy missile attacks and allowed for precise navigation in the featureless desert terrain. With these advantages, the coalition forces were able to quickly conclude the war and thus saved many lives. Space satellite systems will only become more and more vital to homeland defense, weather surveillance, communication, navigation, imaging, and remote sensing for chemicals, fires, and other disasters as time passes.

The ISS is actually a research laboratory in low Earth orbit. Because its design and construction have been contributed by many different partners, it is a symbol of cooperation in space exploration. Former competitors now work together to explore space. Originally, Americans believed that the shuttle would support the ISS, but the Columbia disaster signaled that a shuttle replacement needed to be developed.

The ISS is now supported by mostly expendable rockets and Russian space capsules; however, recently, NASA partnered with private space companies like SpaceX and Boeing for sending humans and cargo into orbit. All space launch systems are being designed taking into account cost reduction and improved dependability, safety, and reliability. Many other countries have launch systems of their own. This means that the private-sector competition will lead to better, safer, and less expensive rockets.

Space Launch

Space launch is the earliest part of a space vehicle's flight (i.e., when the vehicle achieves liftoff and the spacecraft launch system leaves the ground). Note that even when a spacecraft uses multiple stages or different modes of launching, the entire system of the space vehicle from ground to space is considered a single launch system. SpaceX's Dragon spacecraft, for example, lifted off on a Falcon 9 rocket from Space Launch Complex 40 at Cape Canaveral Air Force Station in Florida (**Figure 4.3**). Dragon was carrying more than 5900 lb of research, equipment, and cargo.

There are two main types of launch systems: rocket-based (the current conventional method) and non-rocket-based. The latter includes air-breathing engines or carrier aircraft. There are even a few trans-atmospheric engines being developed—most notably the Skylon, which allows one engine to seamlessly transition from being in an atmosphere to being in a vacuum, i.e., from where the oxidizer is pulled out of the surrounding air to where it must be carried within the spacecraft.

Figure 4.3 A two-stage orbital rocket.

Courtesy of NASA.

Launchers that rely purely on staged rockets are usually at a fixed location on the ground, but they can also be on a floating platform (such as the sea launch vessel developed in the late 1990s). These rockets are currently the only vehicles capable of lifting the heavy payloads required for orbital or inter-orbital missions. The other most popular method for launching vehicles which is used mostly for lighter payloads is an airdrop. This is where a rocket (either single or multistage) is dropped from a carrier aircraft at some height above the ground.

All rockets use the thrust generated by their propulsion system to overcome the weight of the vehicle. Aerodynamic drag differs for different types of rockets. Aerodynamic drag and lift are important forces acting on stomp rockets, bottle rockets, and model rockets. Though aerodynamic drag is significant for air-to-air and ground-to-air missiles, it is not as important for satellite launches because of the different trajectories required to reach orbit.

Pathway to Space

Getting to the ISS, the Moon, Mars, or Venus is not simply a matter of aiming directly at the object and hitting the big red launch button. Because other solar bodies, like our own planet, move in their own elliptical orbits around the Sun, engineers must plan the orbits of spacecraft so that they can arrive at the same location in space/time as their target.

Hitting a target in space is a complicated mathematical challenge similar to shooting a high-flying duck from a racing speedboat. During the Apollo missions, spacecraft were first placed in a parking orbit around Earth before being boosted out of that and into a trajectory toward the Moon.

During the Apollo program launches, the Moon was "the duck," traveling around Earth at an average speed of 3650 km/h. Earth ("the speedboat") was revolving on its own axis as well as orbiting the Sun. The launch point, the Kennedy Space Center in Florida, was moving eastward at 1550 km/h because of Earth's rotation. Earth itself was orbiting the Sun at 106,000 km/h. Moreover, the gravitational fields of the Moon, Sun, and Earth were all tugging at the Apollo spacecraft, which had been launched into space by the powerful Saturn V rocket.

During the Apollo missions, the spacecraft were boosted into Earth-parking orbits where they coasted until they were in the proper positions for beginning the second legs of their journeys. At the scheduled time, the spacecraft were boosted out of the parking orbit into a trajectory or path toward their target, the Moon. These trajectories were and are complicated by the fact that orbital paths rarely form perfect circles. In many cases, the paths of objects around the Sun are elliptical, whose farthest point is called the *apogee*, and the nearest point, the *perigee*.

Unmanned spacecraft have landed on Mars and Venus and have flown past Mercury, Jupiter, Saturn, Uranus, and Neptune. The routes for long journeys into space are worked out using mathematical calculations similar to those used to get a spacecraft to the Moon. They are far more involved, however, because the Sun, the planets, and other objects that occupy our universe, like asteroids and comets, exert their own gravitational and orbital influences on spacecraft that are on extremely lengthy journeys from Earth into space.

Orbital Flight

We will focus on the major aspects of flight for a two-stage orbital rocket launch system. Throughout the flight, the weight of the rocket constantly changes because its propellant is constantly burning. The thrust produced by the engine at launch is greater than the rocket's weight, so the net force accelerates the rocket away from the pad.

A full-scale rocket launch necessarily relies on a sophisticated guidance system to balance and steer the rocket during its flight. The gimballing of the rocket's engines during the flight to produce changes in direction and keep the rocket on target starts the moment the rocket leaves the pad and begins its powered vertical ascent.

The huge thrust and decreasing weight allow the rocket to accelerate upward and very quickly depart from the thick atmosphere surrounding Earth. Although the rocket quickly reaches supersonic speeds, the aerodynamic drag on it is small because of its shape and the lowering air density the further it climbs. As the rocket ascends, it also begins to pitch over and its flight path becomes more horizontal or "parallel" to the surface of the Earth.

Several minutes into the ascent, most spacecraft discard some of their launch vehicle weight in a process called staging. This can be seen in **Figure 4.4**. Once the first stage falls away, the engine on the second, or upper, stage is ignited. The discarded first stage continues on a ballistic flight (i.e., under the influence of gravity) back to Earth. It might be retrieved as the space shuttle solid rocket engines were, or it might be completely discarded as the Apollo Moon rockets were.

Figure 4.4 A two-stage orbital rocket.

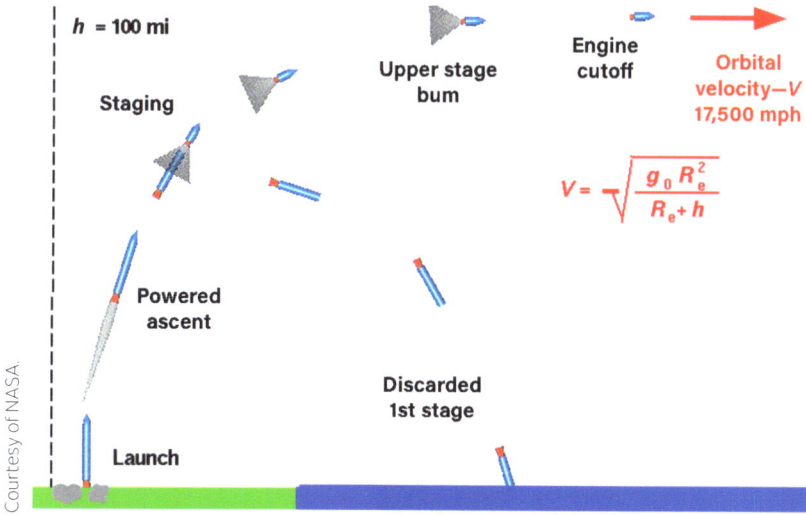

The lighter upper stage continues to accelerate under the power of its engine and continues to pitch over to the horizontal. At a predetermined altitude and speed, the upper stage engine is cut off and its payload is in orbit. The speed required for orbiting Earth depends on the altitude, and it can be calculated by a formula Johannes Kepler developed in the early 1600s:

$$V_2 = g_0 \, Re2 / (Re + h)$$

where

V is the velocity for a circular orbit

g_0 is the surface gravitational constant of the Earth (9.8 m/sec²)

Re is the mean Earth radius (6371 km)

h is the height of the orbit in kilometers

A rocket launched from the Moon or Mars would require a different orbital velocity because of the different planetary radius and gravitational constant. For a 160-km-high orbit around Earth, the orbital velocity is 28,128 km/h.

Re-entry

Launching a spacecraft into space is one thing, but bringing it back is a whole different ball game! Spacecraft re-entry is tricky for several reasons. When an object enters Earth's atmosphere, it experiences two main forces: gravity and drag. It is pulled back to Earth by gravity, but unhindered, it would fall perilously quickly. However, any falling object hits and rubs against the particles of air in Earth's atmosphere, which causes friction. As a result, the object experiences drag (air resistance) and slows down to a safer entry speed. **Figure 4.5** shows an illustration of space shuttle re-entry in terms of high temperature. You can read more about all of this in the article "What if I threw a penny off the Empire State Building?", which you can search online. In the process, friction causes another problem, intense heat. Specifically, the space shuttle faces temperatures of about 1649°C (3000°F) on its lower surfaces. This superheating of the spacecraft surfaces creates unique and challenging problems for spacecraft designers. Blunt-body designs help alleviate the heat problem. On re-entry, a blunt-shaped downward-facing surface creates a shock wave in front of the vehicle that keeps the heat away from it.

Simultaneously, the blunt shape also slows the object's fall by maximizing aerodynamic drag. The Apollo program, which moved several manned ships back and forth from space during the 1960s and 1970s, used blunt-shaped re-entry vehicles and coated them with a special ablative material that burned up upon re-entry and thus absorbed some of the heat. On the other hand, space shuttles are reusable launch vehicles (RLVs). Instead of using ablative material alone, RLVs must incorporate durable insulation that can survive repeated re-entries into Earth's atmosphere.

Figure 4.5 Spacecraft re-entry illustration.

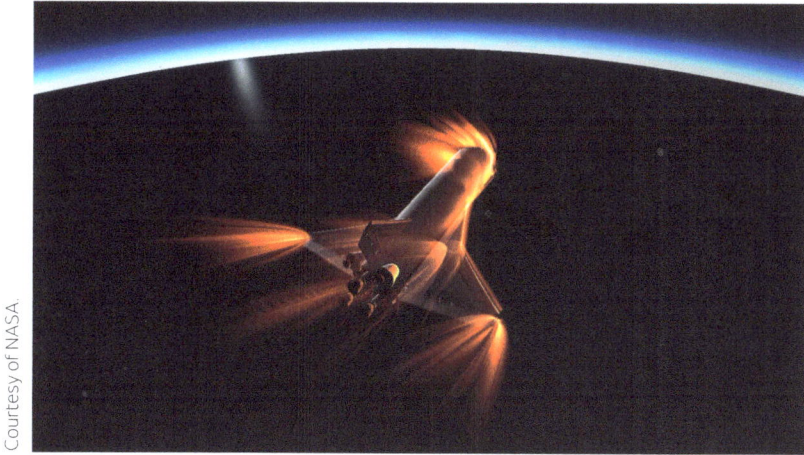

Courtesy of NASA

Satellite

So far, we have focused on vehicles that are meant to come back to Earth—but what about those that are not necessarily coming back? A satellite is any object (i.e., moon, planet, or machine) that orbits a planet or star. So, technically speaking, our Earth is a satellite because it orbits the Sun, and the Moon is a satellite because it orbits Earth.

Usually though, the word satellite is restricted to machines that are launched into space and are meant to stay in orbit around a naturally occurring body such as Earth or another planet. Thousands of these artificial or man-made satellites orbit our planet. They have a myriad of specialized purposes, all of which make our lives on Earth easier or better. For example, **Figure 4.6** shows the Jason Satellite, which monitors global ocean circulation.

Figure 4.6 The Jason-2 satellite orbits Earth.

NASA/JPL-Caltech.

A satellite might be designed to:

- Take pictures of Earth so meteorologists can better predict the weather and track destructive storms.
- Take pictures of other planets, the Sun, black holes, dark matter, or faraway galaxies so that scientists can better understand our solar system and the universe.
- Beam TV signals and phone calls around the world.
- Create a Global Positioning System (GPS). If you have a GPS receiver or a smartphone, these satellites can figure out your exact location.

Why Are Satellites Important?

Because of their bird's-eye view, satellites can see large areas of Earth at one time. This allows them to collect data more quickly than instruments on the ground because they orbit above the atmosphere's clouds, dust, and molecules that block the view from the ground. Satellites can also see into space better than their Earth-based counterparts.

Before satellites, TV signals did not get very far because they could only travel in straight lines. As a result, the signals trailed off into space instead of following the curvature of the planet. Sometimes they would be blocked by mountains or tall buildings.

Phone calls to faraway places were also problematic, since setting up telephone wires over long distances or underwater was difficult and expensive, phone calls would not travel well. Nowadays TV signals and phone calls are sent straight upward to a satellite which, almost instantly, sends the signals back down to different locations on Earth. This capacity allows us to quickly and inexpensively transmit our thoughts from one side of the world to the other.

What Are the Parts of a Satellite?

Satellites come in many shapes and sizes, but most have at least two parts in common: an antenna and a power source. The antenna is used for sending and receiving information—mostly to and from Earth. The power source comprises a solar panel or battery. Solar panels make power by converting sunlight into electricity, whereas batteries can use either chemical processes or even in some cases (like the Voyager) nuclear processes.

Many NASA satellites also carry cameras and scientific sensors. Depending on what they want to gather information about, these instruments might point toward Earth for information about its land, air, and water or toward space for data about the solar system and universe.

Satellite Components

Satellites are versatile machines that vary greatly in complexity, depending on their mission. At their core, all satellites share essential components, including a power source, communication devices, orientation systems, and scientific instruments. These components enable satellites to function reliably in the harsh environment of space. Satellites consist of sophisticated electronic and mechanical components that must withstand the vibrations of a rocket launch and then operate in the environment of space—without maintenance for periods of 15 years or more. Some satellites are simple in design, created for basic tasks such as relaying communications or capturing imagery. For example, **Figure 4.7** illustrates a basic satellite structure with labeled components, including a container, power source, communication device, orientation finder, and scientific instruments. This simple design ensures cost-effective operation while fulfilling specific mission requirements.

Figure 4.7 Components of a simple satellite.

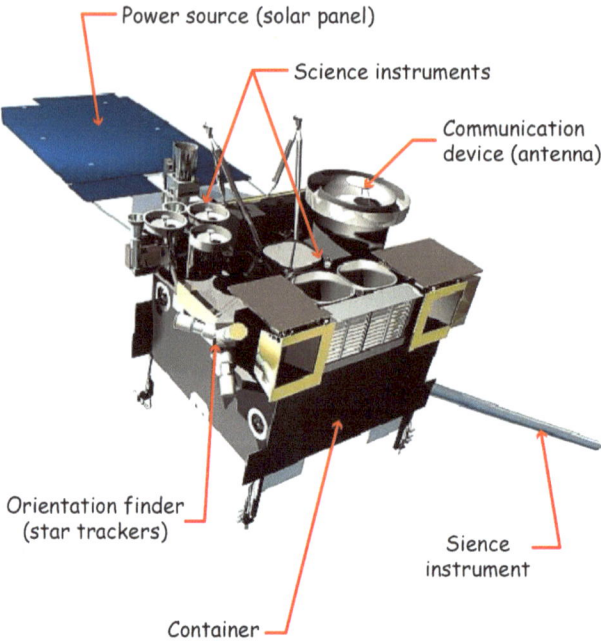

A satellite consists of multiple transponders to communicate with Earth and other satellites. We further explain about satellite communication technologies in Chapter 6. Besides structure, it also contains power, temperature controls, directional thrusters, the communications payload (which receives, amplifies, and retransmits the signals over a designated geographic area).

In contrast, complex satellites like the James Webb Space Telescope (JWST) represent the pinnacle of engineering innovation. Designed for deep-space exploration, JWST features advanced systems, including a gold-coated segmented primary mirror (**Figure 4.8**), which maximizes reflectivity for infrared light. These sophisticated components allow JWST to capture faint astronomical signals and provide insights into the universe's earliest moments.

Figure 4.8 Gold-coated mirrors of the James Webb Space Telescope.

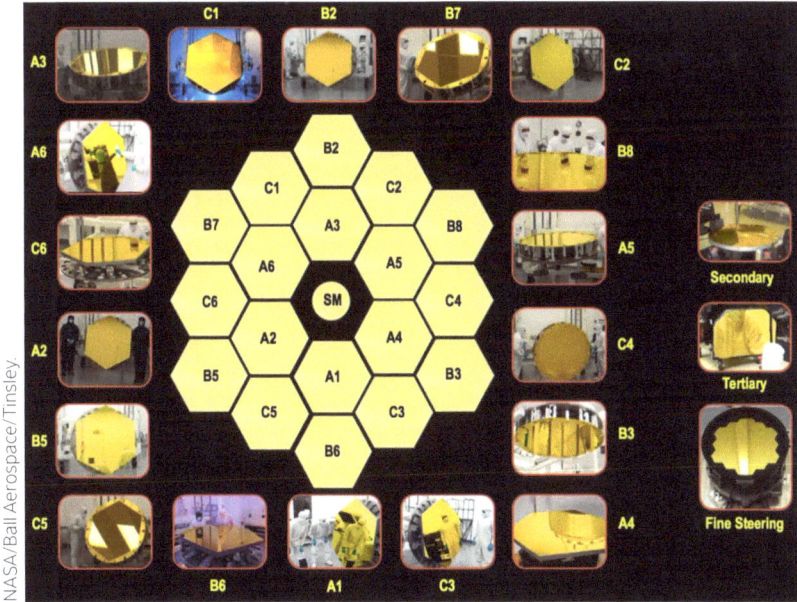

Whether designed for simplicity or complexity, satellites are remarkable tools tailored to meet the unique demands of their missions. Their designs reflect both the ingenuity and resourcefulness of the engineers who created them.

Space, Science, and Technology

Advances in science and technology and our exploration of space have made our world smaller while showing us just how huge the universe really is. Today, a multitude of satellites orbit Earth, monitoring our weather and environment as well as providing global communications.

In the last few years, many new planets have been discovered around the stars, adding further fuel to the question of whether or not we are alone in the universe. At the very least, we have found that the void contains a vast amount of resources that would be extremely useful to us here on planet Earth. Further technological and scientific advancement could even herald a day when we could build an object on Earth from resources mined from another planet.

Perhaps the most exciting benefit of science and technology in space might be the conversion of other planets and/or moons into habitable living spaces for humans. It is even possible that we may discover life on these other planets.

Figure 4.9 illustrates the three major methods the US government uses for searching for extraterrestrial life:

- Searching for extrasolar planets (Kepler spacecraft).
- Listening for extraterrestrial signals that indicate intelligence—out of the scope for this chapter (Allen array).
- Robotic exploration of the solar system (Curiosity Rover on Mars).

Figure 4.9 The three major methods used to search for extraterrestrial life.

NASA/Colby Gutierrez-Craybill/JPL-Caltech.

Astronauts

Astronauts are people who dedicate their lives to space exploration, and keeping them safe is the top priority of spacecraft designers. Spaceships can be manufactured and designed in such a way that even during a catastrophic situation, the astronaut can be ejected or transferred to a safer place.

For instance, NASA's recent Orion program has a safety mechanism that will pull the capsule containing the astronauts away from the rest of the rocket in the event of a catastrophic (i.e., explosive) failure of the lower stages. In orbit, safety zones are created within spacecraft that are more heavily shielded and/or independent than the rest of the station where astronauts can take shelter in the event of unforeseen calamities. On more dangerous missions, humans can be replaced with robots so that critical tasks can be carried out without risking human life.

How to Become an Astronaut?

This is a tricky question, and it is certainly dependent on luck, which is an idea many astronauts may not accept. Of course, the first step is to have a passion for space and not be afraid to talk about your dreams. For example, the path to becoming a NASA astronaut is split into two categories: pilot or mission specialist. Pilot astronauts can serve as commanders or pilots of missions and are responsible for the vehicle and crew. There are also many different requirements to fulfill. Mission control applicants must be able to pass a space physical, including the following items:

- Distance visual acuity: 20/200 or better uncorrected, correctable to 20/20, each eye.
- Blood pressure: 140/90 measured in a sitting position.
- Height between 58.5 and 76 in.

Many current astronauts recommend to lay out a plan if you really want to be an astronaut. Cady Coleman, a current astronaut said about her experience in this article[1]: "The biggest challenge about being involved in the space program is the need to be able to be good at and know a lot about a lot of things," Coleman says. "It's not just chemistry anymore."

Space Technology—Useful for Humans on Earth

The number of everyday objects and technologies that have been developed or enhanced as a direct result of space exploration is astonishing. More than 30,000 space-derived innovations are used in our daily lives, such as laser surgery, brain scanners, and water purifiers. Even the equipment that allows doctors on Earth to monitor an astronaut's life signs has been applied to life support machines in intensive care units. Without space exploration, many of the things that make our modern lives possible might not have been discovered, such as baby food and vacuum cleaners.

Future Space Projects

Space is always changing, and new and innovative ideas are always being put forth to get humanity to the stars. One of these ideas is the Google Lunar X-Prize, an extension of the famous X-Prize that gave us the first commercial spaceflight. This prize is about changing the game when it comes to lunar exploration by encouraging the growth of the private sector in the space industry. Getting to the Moon has been anything but cost-effective—so far. But with a $30 million prize designed to inspire pioneers to do robotic space transport on a budget, that all could change very quickly. Teams from around the world are competing to land a robot safely on the Moon, move 500 m on, above, or below the Moon's surface, and send HDTV Moon casts back to Earth. The prize deadline was extended, but the contest was officially closed in 2018 without a winner.

---

1 "HSF," Spaceflight.nasa.gov, n.p., 2019, Web, March 31, 2019.

Energy from Space—Used on Earth

Two of the major drawbacks of solar energy plants on Earth are their requirement for large space and their inability to generate power during the nighttime or on cloudy days. Unfortunately, on Earth, there are no ways around these issues; however, in space, there is no atmosphere that can produce cloudy days, and satellites can be placed into geostationary orbits and bounce sunlight off each other so they are always receiving energy—energy that can be beamed down to Earth and used any time of the day.

A space-based solar power (SBSP) station would use an array of mirrors to concentrate the Sun's rays on photovoltaic cells to produce electricity, which would then be converted into a microwave beam and directed at an antenna on Earth where it would be converted back into electricity and fed into the grid.

Being able to use renewable energy for the production of existing fuels in a carbon-neutral cycle would be a total game-changer and thus warrants significant attention. The primary environmental benefit of SBSP stations is in the form of nearly carbon-free renewable energy. One of the places where you can read more about this is the *Electric Space*[2] book.

Wormholes

One of the most difficult challenges to interstellar flight is the vast amount of time it would take to cover the enormous distances between stars. An artist has created a wormhole in **Figure 4.10**. Wormholes, like those popularized in the popular sci-fi movie *Interstellar*, have the potential to circumvent the speed of light. The speed of light is the most limiting force for interstellar travel. Since nothing can travel faster than light, any potential vessel that travels via conventional means would take several thousands, if not millions, of years to travel from Earth to the nearest star systems.

---

2 *Electric Space: Space-Based Solar Power Technologies & Applications* by D.R. Jones and A. Baghchehsara.

Figure 4.10 A science fiction look at a spacecraft traveling through a wormhole.

Digital art by Les Bossinas.

However, if wormholes do exist, they could act like conduits connecting one region of space to another. A space traveler might be able to pass through the conduit instantaneously. Even though we know that Einstein's special relativity means that objects cannot travel faster than light, we also know that space-time itself can be distorted.

Wormholes would allow us to find a loophole in physics despite the fact that it would take an almost impossible amount of energy to distort space-time enough to move a starship. The keyword here is "almost." Wormholes are at the very cutting edge of theoretical physics, but humans have already begun researching potential spaceship configurations based on what is known as the *Alcubierre* drive, or the *Alcubierre* warp drive, which was proposed by the Mexican theoretical physicist Miguel Alcubierre.

Preparation to Live in Outer Space

One of the hottest topics right now is about setting up mining operations on solar bodies. Companies like Planetary Resources want to set up mining operations on asteroids, and national governments like China are researching ways to mine large bodies like the Moon for valuable Helium-3. Many of these resources could be shipped back to Earth or used in space to construct even more advanced spacecraft.

The Mars One Project

The Mars One organization was founded in 2011 and developed a strategic plan for landing people on Mars and leaving them there for a permanent settlement. **Figure 4.11** shows the Mars colony designs beside each other. Mars One's stated goal is to establish a human settlement on Mars, the Red Planet, with the first crew landing in 2025. The company is a non-profit foundation that plans to establish a permanent Martian human settlement using existing technology.

Figure 4.11 Mars One colony.

3000ad/Shutterstock.com.

Mars One's mission plan integrates components that are well tested and readily available from industry leaders worldwide. It calls for an initial launch of cargo missions, followed by preparation of a habitable settlement by unmanned means and finally by human landings. The most interesting and perhaps most controversial part of the plan is that Mars One is not accounting for a return trip. All settlers who make the journey will do so knowing that they will probably never be able to return to Earth. Despite this, more than 200,000 registered for the first selection program when it opened in 2013.

3D Printing in Space

One of the most exciting advances in recent years has been the mass consumerization of 3D printing technologies. These technologies allow users to bring a robotic manufacturing device into their home that produces a near-infinite replication of arbitrarily complex parts with an astounding degree of accuracy. It is accomplished by using a machine head on a gantry to layer successive plastic layers on top of one another.

Many have asked what this new movement means for the future of manufacturing here on Earth—and others have wondered what it means for our lives beyond our home on Earth. Perhaps the most futuristic use seen so far was when NASA "emailed" a replacement wrench to space to be printed on a made-in-space 3D printer stationed on board the ISS. The resulting product is a functioning adjustable tool that astronauts can use around the station.

To put this into perspective, it currently costs $322 per kilogram to ship an object to space on board a Falcon Heavy rocket. To ship a metal wrench weighing 0.4 kg from Harbor Freight would cost at least $249, which does not include the weight and space cost of any packaging material.

Now, granted the wrench bought from the hardware store is probably metal, the printer on the ISS is constructed out of plastic. Metal printers do exist, and they are starting to come down in price and size so they can be used in home (and eventually in space). However, the real benefit of having a 3D printer in space is its non-specific nature. This means that with just a printer and raw material, almost any component (that can be replicated out of plastic) can be built in space—negating the current issue of astronauts having to wait for any replacement part until the next rocket is available to launch.

Historically, this produce on Earth ship to space model has been the only effective way of producing the parts needed to make living in space possible. Fabrication of complex components usually requires big machines and is typical of subtractive construction methods, lots of dust, waste, energy, and raw material. Since 3D printers use an additive construction method, there is little to no waste produced by the process—no dust, sparks, etc. Virtually, the only waste products are fumes produced from melting the plastic and plastic rafts used to adhere the printed part to the print bed. Compare this to traditional manufacturing processes with dust and debris flying everywhere, as material is cut away from an existing hunk of material. Can you imagine the chaos that would be unleashed by using a belt grinder in a zero-gravity environment?

In the near future, we could be sending only hunks of raw material up to space rather than complete parts. Three-dimensional printers offer the capacity for producing infinitely complex parts, performing infinite replications, and embodying the skills of dozens of different profes-sional craftspeople. This allows us to not only start replacing current parts with 3D-printed substitutes, but design parts to be manufactured solely in space. Astronauts could repair devices on an as-needed basis, and we could finally start producing items in space like the building blocks of colony ships.

Currently, this is a theoretical concept about the application of robotic manufacturing to space travel, in an effort that we can overcome one of the most limiting characteristics of our current space transportation paradigm. The robotosphere would essentially be a robotic ecosystem that sustained itself beyond our atmospheric boundary. Robots would mine resources and ship them, while others would process and build with them. Still, others would repair the builders. In time, we could see this ecosystem extend and sustain itself almost entirely without human intervention. Schematics of satellites and starships could be uploaded to orbiting factory satellites and produced on demand for astronaut residents.

If we can construct a simulation reasonably quickly and if it is accurate, it is possible that future designers could input data about the environment where the object will be used (such as atmospheric pressure and gravity) and have a simulated world created around their design. This would allow for simulated real-world testing. Three-dimensional modeling software would then become more intuitive for a designer or an engineer. Users would be able to simply tug and pull on their design in a virtual space to see how it reacts instead of having to visualize material behavior in their heads.

Conclusion

Newton once said about space exploration: we all should work together as one to achieve greater heights in this particular field of search, without difference of status, jobs, color, and nation.

There are few instances in the breadth of human endeavor where we have so much to gain and so obviously need to set aside our differences, as there are with space travel and exploration. While we have already reaped the rewards of past efforts, it is now time for us to press onward and find the internal strength to accomplish the great feats necessary to take our civilization to the stars.

NASA has helped many aspiring young scientists in fulfilling their dreams and goals, but it will be up to the next generation of scientists, designers, artists, and engineers to create a lasting presence for humanity beyond the confines of our Earth. With the unprecedented speed at which our society is advancing and the increasing presence of private companies in space and space exploration activities, the time has never been better to make the leap into the void.

Chapter 05

Orientation and Control

Figure 5.1 The two major types of flight—atmospheric and space.

In this chapter, we will start on a new topic in aeronautical engineering: flight mechanics. We will also provide an overview of spacecraft controllability, suborbital flight profiles, orbital flight vehicles, and atmospheric flight vehicles like aircraft.

Laws of Motion

A simplified review of the three laws of motion was proposed by Sir Isaac Newton in 1665, and they help explain how an object flies and reaches its destination.

First Law of Motion—the Law of Inertia

An object at rest will remain at rest unless acted on by an unbalanced force, and an object in motion continues in motion with the same speed and in the same direction unless acted upon by an unbalanced force. In other words, there is a natural tendency of objects to keep on doing what they are doing. All objects resist any change to their state of motion.

Second Law of Motion

Acceleration is produced when a force acts on a mass. The greater the mass of the object being accelerated, the greater the amount of force needed to accelerate it. In other words, heavier objects require more force to move the same distance than lighter objects, which is something everyone unconsciously knows. However, the second law goes further than that by providing the exact relationship between force (F), mass (M), and acceleration (A) in the mathematical equation:

$$F = M \times A.$$

Third Law of Motion

When an object is pushed in one direction, there is always a resistance of the same size in the opposite direction. In fact, for every action, there is an equal and opposite reaction. In other words, for every force, there is a reaction force that is equal in size, but opposite in direction.

Whenever an object pushes another object, it gets equally pushed back in the opposite direction.

How does this work for a rocket? The force of its powerful engines acts to push down on the ground, and the reaction from the ground pushes the rocket upward with an equal force (**Figure 5.2**). In the next section, we will derive some equations for a straightforward cruise based on this rule.

Figure 5.2 Newton's third law of motion demonstrated on a rocket.

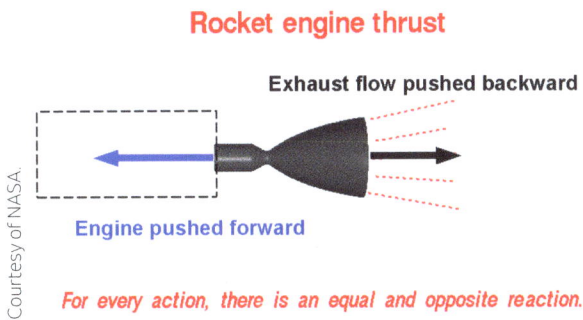

Rocket engine thrust

Exhaust flow pushed backward

Engine pushed forward

For every action, there is an equal and opposite reaction.

Courtesy of NASA.

Orbits

An orbit is a path around an object in space; for instance, the Moon is said to orbit around Earth. You are in orbit right now! We all are! That is because Earth is orbiting around the Sun. The ISS orbits Earth. An object in orbit is called a satellite.

The path of an orbit is curved somewhere between a circle and an oval, but technically speaking, it is called an ellipse. A comet, for example, follows a very long and thin ellipse. Sometimes the comet's ellipse is close to the Sun and moving quickly, but mostly, it is very distant from the Sun and moving slowly. The Moon's orbit, on the other hand, is almost circular.

Why are there these differences? To find out, we will look at the three laws about the Sun and planets developed by Johannes Kepler (1571–1630).

1. The path of the planets about the Sun is elliptical in shape, with the center of the Sun being located at one focus (the law of ellipses).

2. An imaginary line drawn from the center of the Sun to the center of a planet will sweep out equal areas in equal intervals of time (the law of equal areas).

3. The ratio of the squares of the periods of any two planets is equal to the ratio of the cubes of their average distances from the Sun (the law of harmonies).

Kepler was able to develop these three laws describing the motion of planets in a Sun-centered solar system after studying the data that had been carefully collected by his mentor, Tycho Brahe. Modern science no longer accepts Kepler's explanations of the underlying reasons for these orbits, but the actual laws themselves are still accepted as accurate descriptions of what happens.

You now know from Newton's first law of motion that a moving object will continue moving unless something pushes or pulls on it. For example, a satellite would fly off into space without gravity, and with gravity, it is constantly pulled back toward the Earth. This tug of war between its velocity and gravity keeps the satellite in orbit (**Figure 5.3**).

Figure 5.3 The tug of war between gravity (*o*) and velocity (*v*) keeping the Moon in orbit around Earth.

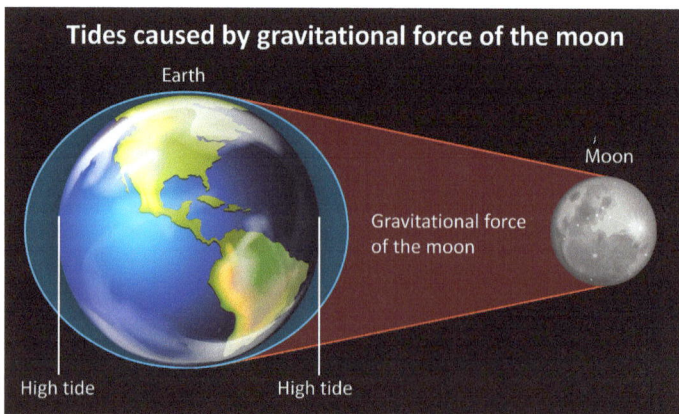

Tides caused by gravitational force of the moon

Earth

Moon

Gravitational force of the moon

High tide   High tide

BlueRingMedia/Shutterstock.com.

Height is how far up something is. Interestingly, objects at different heights orbit at different speeds (**Figure 5.4**). The ISS is about 320 km above Earth and must move about 800 km/h; this takes about 90 min to orbit once around Earth. The Moon is much higher at about 400,000 km from Earth. So the Moon must move about 3500 km/h and takes about 28 days to orbit once around Earth. Earth takes a year to orbit the Sun once, but Pluto takes about 248 Earth years to orbit the Sun once.

Figure 5.4 Meteor glowing as it enters the Earth's atmosphere.

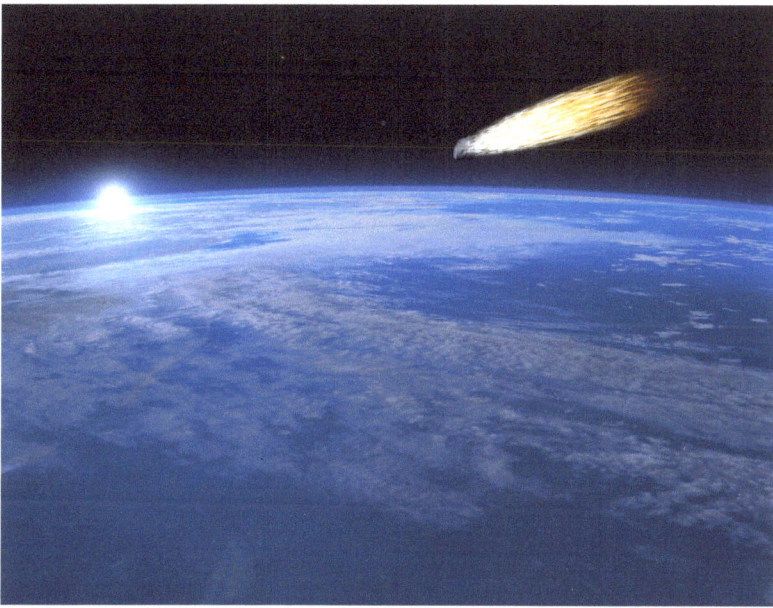

Elenarts/Shutterstock.com

Moving in Orbits

When a satellite's speed is balanced by the pull of the Earth's gravity, it will orbit Earth, but without the speed and gravity balancing each other out, it will either fly off into space in a straight line or fall back to Earth. Satellites orbit Earth at different heights, different speeds, and along different paths. The most common types of orbit are the geostationary and the polar.

Figure 5.5 Principal orbits for artificial satellites around the Earth.

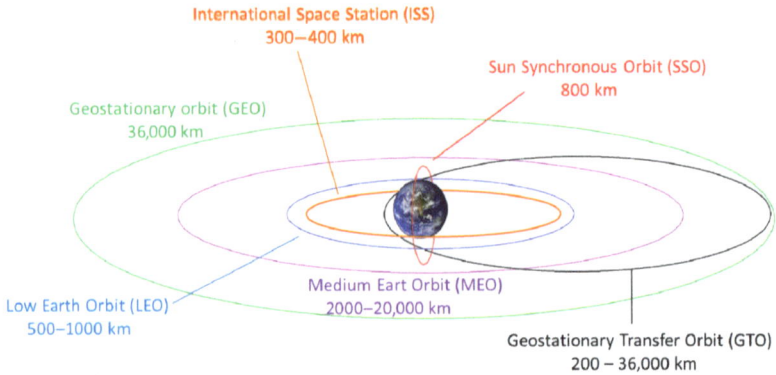

International Space Station (ISS)
300–400 km

Sun Synchronous Orbit (SSO)
800 km

Geostationary orbit (GEO)
36,000 km

Low Earth Orbit (LEO)
500–1000 km

Medium Eart Orbit (MEO)
2000–20,000 km

Geostationary Transfer Orbit (GTO)
200 – 36,000 km

Reprinted from Cottin, H., Kotler, J.M., Billi, D. et al. Space as a Tool for Astrobiology:
Review and Recommendations for Experimentations in Earth Orbit and Beyond.
Licensed under CCBY 4.0, https://creativecommons.org/licenses/by/4.0/.

A geostationary satellite travels level with the equator from west to east in the same direction as the spinning Earth and at the same rate (**Figure 5.5**). Consequently, from Earth, it looks stationary since it always sits above the same location.

As demonstrated in **Figure 5.6**, a *polar-orbiting satellite* travels from pole to pole or from north to south to north again. It can scan the entire globe, one strip at a time, while Earth spins underneath it.

Figure 5.6 Leaving a parking orbit.

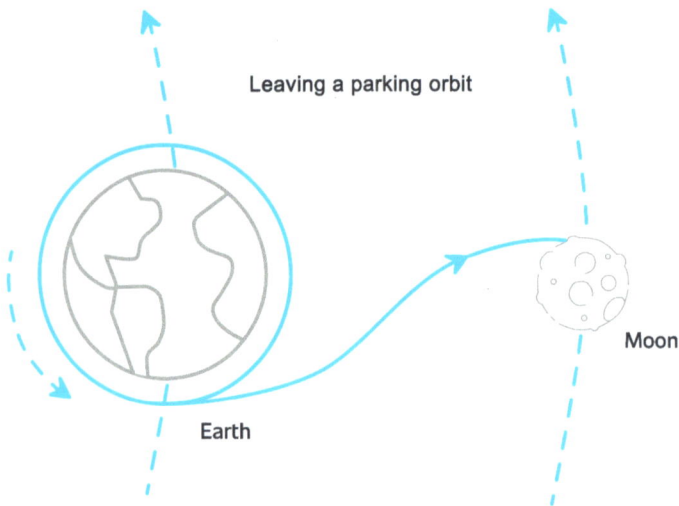

Leaving a parking orbit

Moon

Earth

© SAE International.

Why Do Satellites Not Crash into Each Other? Actually, they can. Because satellites in space are monitored by NASA and other US and international organizations, collisions are fortunately rare. Moreover, each satellite is launched into an orbit designed to avoid other satellites. However, orbits can change over time, and the chances of a crash increase as more and more satellites are launched into space and left to stay there for years at a time. The first and so far only accidental collision between two man-made satellites occurred in February 2009 when two satellites, one American and one Russian, collided high above the planet. Both were totally destroyed.

Gravity Assist

Comets and other bodies in solar orbit naturally experience changes in their orbits when they pass close by a planet or a moon. Sometimes the path and speed of a spacecraft are altered by using the relative movement and gravity of a planet or other astronomical object. This is called a gravitational slingshot, a gravity assist maneuver, or a swing-by. It is usually used to save propellant, time, and expense.

Several robotic spacecraft have used this maneuver so they could reach their targets much further away from the Sun. *Voyager 2* was launched in August 1977 and flew by Jupiter for reconnaissance and for a gravity assist boost so it could reach Saturn. *Voyager 1* was launched the following month, used the same technique, and reached Jupiter before *Voyager 2*. *Voyager 2* then obtained a gravity assist boost from Saturn and another one later from Uranus to get all the way to Neptune and beyond. *Galileo* got a gravity assist boost from Venus and two from Earth while orbiting the Sun on the way to its destination, Jupiter. Cassini took two gravity assist boosts from Venus, one from Earth, and another one from Jupiter to gain enough momentum to reach Saturn.

The gravity assist technique usefully adds or subtracts momentum to a spacecraft's orbit. It has generally been used in solar orbit to increase a spacecraft's velocity and propel it outward in the solar system, much farther away from the Sun than its launch vehicle would have been capable of. Since a flyby can also decrease a spacecraft's orbital momentum, *Galileo* decreased its energy relative to Jupiter with a gravity assist flyby from the Jovian moon Io. This way, it was possible to decrease the mass of rocket propellant needed for insertion into a Jupiter orbit.

The two *Voyager* spacecraft provide a classic example (**Figure 5.7**). They were launched from a Titan-III/Centaur, destined for Saturn and beyond. But they could only reach Jupiter, which is halfway to Saturn. If Jupiter had not been there at the right time, they would have reached aphelion, the farthest orbital point near Jupiter's orbital distance of about 5 AU (i.e., five astronomical units or 750,000,000 km) from the Sun. Their perihelion or the nearest orbital point would have been about Earth's orbital distance of 1 AU (150,000,000 km from the Sun). They would have remained in that orbit until a planet or something else caused it to change.

Figure 5.7 Flight profiles for *Voyager 1* and *Voyager 2* to Saturn and beyond using gravity boosts.

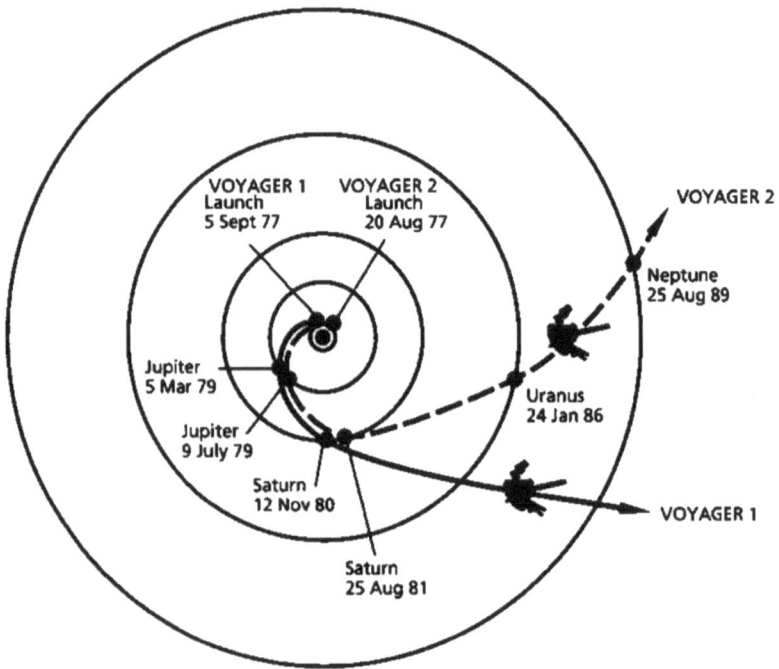

But of course their launch times were planned so that Jupiter coasted by at just the right time. The spacecraft felt Jupiter's gravity and started falling toward it. The spacecraft's velocity brought it close behind Jupiter in its solar orbit, but not close enough for impact. As *Voyager* traveled away from Jupiter, it slowed down again with respect to Jupiter and eventually reached the same speed it had had on its way in.

The spacecraft approaching and departing from Jupiter is like a bicyclist speeding up while going downhill into a valley and then slowing down again for the uphill road leaving the valley.

Figure 5.8 is a vector diagram showing the situation in two simplified dimensions: magnitude of velocity and direction of velocity. The magnitude and direction of the spacecraft's trajectory on its way in toward Jupiter are shown in the lower right. Its magnitude and direction on its way out away from Jupiter are shown in the upper left. You can see that the direction of the spacecraft's velocity has changed because of the accelerating force of Jupiter's gravitation, but not its magnitude.

Figure 5.8 Vector diagram of *Voyager*'s gravity boost from Jupiter toward Saturn.

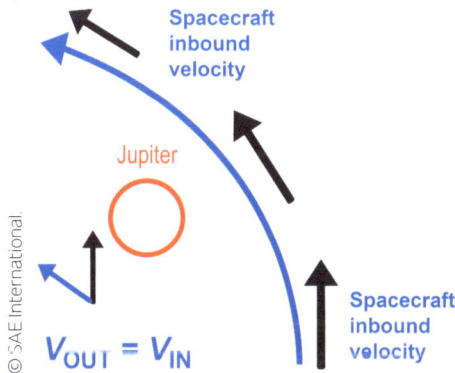

Note that V_{IN} and V_{OUT} represent velocity before and after being noticeably changed by Jupiter's presence. Near the middle of the diagram, the long arrow adjacent to Jupiter indicates where there's a significant but temporary increase in the magnitude or speed as its orbital direction changes. Note that these speeds are all with respect to Jupiter.

Let us look at this in terms of the cyclist again but ignoring air friction, vehicle friction, etc., as they are virtually absent in the spacecraft's situation. V_{IN} shows the cyclist approaching a downhill grade into the valley. V_{OUT} shows that the cyclist has slowed down again at the top of the ensuing uphill grade. Indeed, after going through the valley, the cyclist's direction might have changed, but in the end, there's been no lasting change in speed.

However, for someone watching this from beyond Jupiter's perspective, something else has happened as well, which can be seen in the cartoon in **Figure 5.9**. In the cartoon, the child sees his tennis ball moving away from him at 50 km/h. So does the Sun sitting on the station platform. The engineer driving the train sees the ball coming at 100 km/h since the train is moving at 50 km/h with respect to the ground. The train and ball interact at 100 km/h.

Figure 5.9 The assisted space ball.

Courtsey of NASA. Concept by Charles Kohlhase, based on artwork by Gary Hovland.

The ball rebounds from the front of the train at nearly the same 100 km/h, but this can be added to the 50 km/h speed of the train, which it acquired from the train. The result seen by a stationary observer approaches a total of 150 km/h. In the same way, the velocity of a spacecraft is added to the velocity of the massive speeding planet so it "rebounds" with an even higher velocity. In the case of the spacecraft though, it is a gravitational rather than a mechanical interaction that causes this to happen.

Suborbital Flight Mechanics

In outer space, the suborbital flight mechanics equation derivations and the basics are the same as they are for atmospheric flight. The only difference between these is that in outer space, there is no atmosphere and no significant drag above 24,500 m. Flight profiles are similar too, but during the planned, not executed Lynx Mark 1 (XCOR) flight, participants would have experienced an exhilarating rocket ride to space, with out-of-this-world views and the feeling of weightlessness. In the daytime, the sky would have been black, and Earth's curvature along with the thin blue line of the atmosphere would have been visible. After re-entry, participants would have continued to enjoy this view until they descended to lower altitudes before landing on the same runway where the space plane had taken off.

When the California-based company XCOR Aerospace was in the race competing in the emerging suborbital spaceflight market, they were developing the XCOR Lynx, a suborbital horizontal-takeoff, horizontal-landing (HTHL), rocket-powered spacecraft. It was planned that the Lynx would carry one pilot, one ticketed passenger, and/or a payload above 100 km altitude. In March 2014, the passenger ticket was projected to cost $95,000.

At the time, the XCOR Lynx Space Flight vehicle's stats were remarkable. It was planned to break the sound barrier and fly at supersonic speeds 58 sec after taking off. Immediately after takeoff, it would have climbed at a 75° angle with a constant push of 3Gs. Its maximum speed during flight to space would have been Mach 2.9 or 3550 km/h (2200 mph). After 3 min of rocket-like acceleration, the Lynx spacecraft would reach 58 km and then switch off its engine. It would still have had enough speed to reach the maximum of 103 km (338,000 ft) in about another 1.5 min. The view the passengers would have had from space would have been absolutely stunning. As the astronauts say, it is that view back to Earth that changes lives, calling it "the overview effect."

The Lynx flights were to start in 2014 from Curacao in the Caribbean and the Mojave Desert in the US; however, XCOR Aerospace changed its plans and finally in 2017 it went bankrupt.

Future of Manned Suborbital Spaceflight

Private companies such as Virgin Galactic, XCOR, Armadillo Aerospace, Airbus, Blue Origin, and Masten Space Systems continue to take an interest in suborbital spaceflight. NASA and others continue to experiment with scramjet-based hypersonic aircraft, and they might eventually be used with flight profiles for suborbital spaceflight. Some non-profit groups like ARCA SPACE in New Mexico, US, and Copenhagen Suborbital in Denmark are also attempting rocket-based launches.

Air Flight Mechanic

Flights for general aviation and air transportation typically follow the flight profile shown in **Figure 5.10** and explained further below.

- **Preflight:** This portion of the flight occurs on the ground, where flight checks, push-back from the gate, and taxiing to the runway take place. You take your seat, stow your hand luggage, and do up your seat belt.
- **Takeoff:** The aircraft is powered up and accelerates down the runway.
- **Departure:** The plane becomes airborne and ascends to a cruising altitude.
- **Cruise:** The aircraft travels through one or more airspace centers or zones and nears the destination airport. It spends most of the duration of the flight during cruise.

Figure 5.10 Flight profile for general aviation.

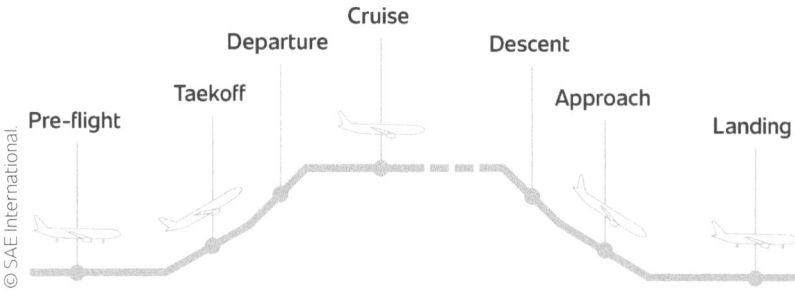

At this point in the flight profile, we will derive equations for a normal cruise flight with a zero angle of attack, which means the aircraft is flying in the same direction as the movement of the air at steady state or at a constant speed. It is very easy, but do you have any idea about how to start? Yes, of course—Newton's third law of motion is the key, and it is very simple.

$$\sum F = M \times A.$$

When you are reminded of laws of motion, do not take another step before you draw a free body diagram, i.e., a diagram that shows the relative magnitude and direction of all the forces acting upon the object in the given situation. A cruise flight as you know has no angle of attack, so **Figure 5.11** shows its free body diagram.

Figure 5.11 Free body diagram of a cruise flight.

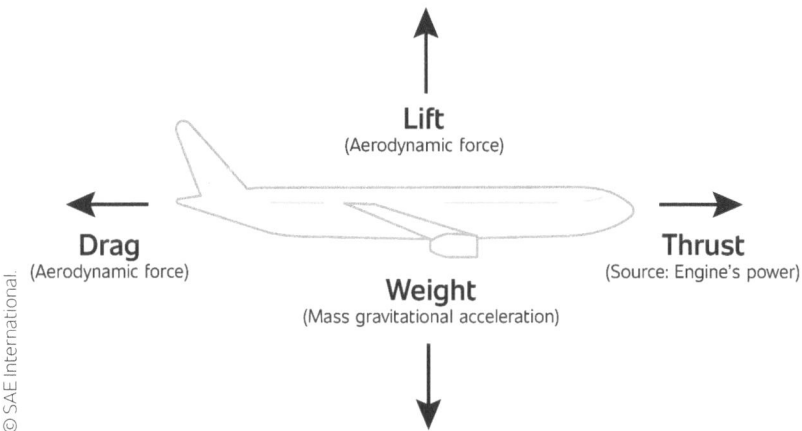

Once you have done the free body diagram, you are almost there.

$$\sum Fx = M \times A$$

$$\sum Fy = M \times A.$$

The forces are divided into two vectors X and Y, so

$$\text{Thrust} - \text{Drag} = M \times A = 0 \quad (\text{i.e., there's no acceleration})$$

$$\text{Lift} - \text{Weight} = M \times A = 0 \quad (\text{i.e., there's no acceleration})$$

So lift is equal to weight and drag is equal to thrust.

If you are motivated to do so, you can start deriving equations for the climb approach and any other situation. However, that is beyond the scope of this book.

- **Descent:** The aircraft descends and is maneuvered to the destination airport.
- **Approach:** The aircraft is aligned with the designated landing runway.
- **Landing:** The aircraft lands on the designated runway, brakes, slows down, taxis to the destination gate, and parks; then, the flight is safely finished. There is something missing in all this, is there not? We are saying the aircraft flies from an airport, takes off, cruises, goes to another airport, descends, approaches, and lands, but do you know how the aircraft really controls itself where there is no asphalt surface for it to drive on? It is time now to delve into aircraft controllability (**Figure 5.12**).

Figure 5.12 A Piper PA-28 Cherokee landing sequence at the COPA Convention at Wetaskiwin, Alberta, Canada.

Photo Credit: Ahunt.

Aircraft Control

We will start with the aircraft. Let us suppose you are flying within the US, perhaps from Los Angeles to San Diego. Your flight is halfway there, and you are cruising at flight level 330, which is at about 10 km altitude (33,000 ft). Now imagine that your airplane encounters a technical problem and all engines suddenly stop working. Though we are hypothesizing, this could actually occur if the fuel ran out or if we flew through a cloud of volcanic ash. In 1998, British Airways Flight 9, a Boeing 747, flew through a cloud of volcanic ash above Indonesia, which resulted in the failure of all four engines. In this situation, there is no other option but to start a gliding flight.

How far do you think gliding aircraft can glide? A typical commercial aircraft can glide more than 170 km from 10 km altitude. Thus, in this specific scenario, the pilot could actually choose to glide either to LA or to SD safely. Now that is impressive, is it not! Do not worry if your guess was incorrect. A solid understanding of flight mechanics can allow you to make this kind of calculation using some basic knowledge about the aircraft.

Depending on the type of aircraft flown, there are different ways for controlling its direction, altitude, and the general controllability of its flight. You might be surprised to know that although electronic control devices are advanced today, mechanical cables and rods are used to transmit the forces from the cockpit to the aircraft's control surfaces.

Aircraft Cockpit

The pilot controls the aircraft from the cockpit, also known as the flight deck, which is usually near the front of the aircraft and is enclosed these days, except on some small aircraft (**Figure 5.13**).

Figure 5.13 Cockpit of the Bombardier Q400 NextGen aircraft.

Photo Credit: Rick Rydell.

This is the place where the instrument panel is, which contains the flight instruments. The controls that enable the pilot to fly the aircraft are there too. The cockpit is a wonderful and exciting place to visit!

You may ask the flight crew if you can visit it at a suitable time, and then, you will understand that a simple flight has to be very organized and that it is not always an easy job. At first sight, you will notice that there are more than a hundred control systems for managing a flight.[1]

---

1 http://www.bombardier.com/en/corporate/media-centre/multimedia-library/details?categoryID=L3-Bombardier%20Q400%20NextGen

Aircraft Control Surfaces

Since aircraft function in a three-dimensional world, it is necessary to control their attitude or orientation in flight. For the purpose of analysis, a flying aircraft is thought of as a particle. It has six degrees of freedom; it can move independently in six possible ways in three-dimensional space. It can move along three perpendicular axes: forward, backward, up, down, left, and right. It can also rotate about those same perpendicular axes: pitch, yaw, and roll (**Figure 5.14**).

Figure 5.14 Illustration of pitch, yaw, and roll.

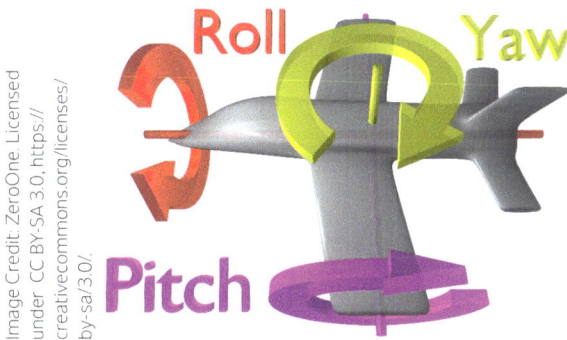

Image Credit: ZeroOne. Licensed under CC BY-SA 3.0, https:// creativecommons.org/licenses/ by-sa/3.0/.

An aircraft's center of gravity is the point where its average mass is located. It is useful to define a three-dimensional coordinate system through this point with each axis perpendicular to the other two. Using this, we can then define the aircraft's attitude by the amount of rotation along these principal axes. Rotation about the lateral axis is called pitch, rotation about the longitudinal axis is called roll, and rotation about the vertical axis is called yaw.

Have you looked at an aircraft's moving surfaces carefully? Do you know what they are called and what they do? **Figure 5.15** shows a typical aircraft along with its detailed control surfaces.

Figure 5.15 Aircraft control surfaces.

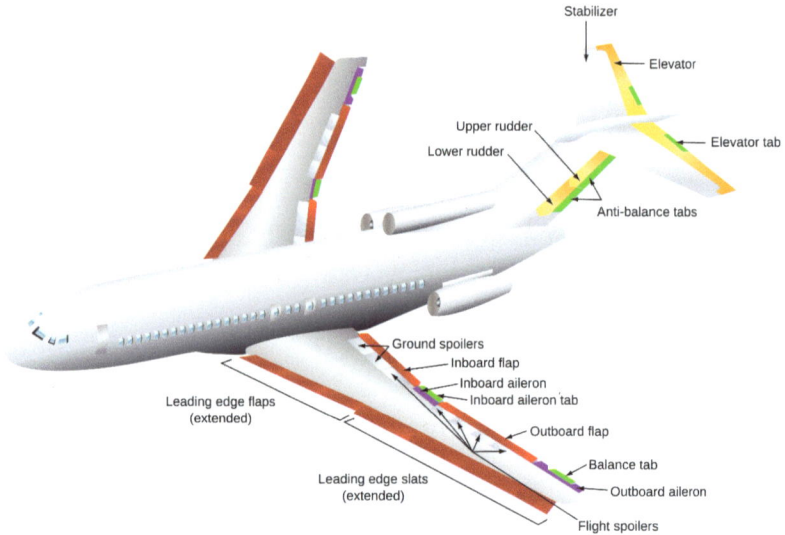

Image Credit: Federal Aviation Administration.

Earlier, you learned how an airfoil creates forces. Smaller airfoils such as elevators, rudders, ailerons, and flaps are the control surfaces used to control aircraft. These are the basic means for making one fly smoothly and go wherever they want. There are other control surfaces, but these four are enough for the basics.

Ailerons primarily control *roll*. The left aileron goes up and the right aileron goes down, if the pilot moves the control stick to the left or turns the wheel counter-clockwise. Lift on a wing is reduced if its aileron is raised and increased if its aileron is lowered. As a result, moving the control stick to the left causes the left wing to drop and the right wing to rise.

The *elevators* primarily control *pitch*. These are the moveable parts at the back of the horizontal stabilizer on the aircraft's tail, and they move up and down together. The elevators go up if the pilot pulls the control stick backward and down if the control stick is pushed forward. Raising the elevators pushes the tail down and causes the nose to pitch up. The aircraft's wings will then fly at a higher angle of attack, which in turn generates more lift and also more drag.

The *rudder* primarily controls *yaw*. It is part of the aircraft's tail assembly and is usually mounted on the trailing edge of the vertical stabilizer. The rudder turns to the left if the pilot pushes the left pedal and to the right if the pilot pushes the right pedal. A right-turning rudder pushes the tail left and causes the nose to yaw to the right. If the rudder pedals are centered, the rudder returns to neutral and stops the yaw (**Figure 5.16**).

Figure 5.16 KLM Fokker 70, showing the deployed positions of both flap and lift flight controls.

Photo Credit: Arpingstone.

After the aileron, elevator, and rudder, the *flap* is the next important surface that helps to control an aircraft. It raises the maximum lift. If the flaps are deployed, they deflect down, thereby increasing the effective curvature of the wing. They are generally used during low-speed, high-angle-of-attack flights, especially at takeoff and descent for landing. Although high-lift flaps are not used as ailerons at all on commercial aircraft, sometimes they function primarily as ailerons. On some aircraft, ailerons will "droop" when flaps are deployed, thus acting as both a flap and a roll-control onboard aileron.

These four are the primary surfaces for controlling an aircraft, but there are others that help the flight go as expected too: spoilers, air brakes, slats, and trim tail. In a helicopter, it is a completely different ball game, which we learn next.

Helicopter Control

Unlike an airplane that forces air over a pair of fixed wings to fly, a helicopter does so by spinning a rotor blade at high speeds. They are highly maneuverable aircraft whose invention is attributed to Leonardo da Vinci (1452–1519). Russian-born Igor Sikorsky (1889–1972) developed the first practical design in the 1930s (**Figure 5.17**).

Figure 5.17 Commodore Frank Erickson, Dr. Igor Sikorsky, Sikorsky Helicopter in 1942—HNS-1 C.G.

Photo Credit: U.S. Coast Guard.

Typically, helicopters are used these days for military transportation and air–sea rescue. They fly upward against the force of gravity by using their rotors to throw air down beneath them. Each blade in a helicopter's rotor is an airfoil just as the wings of an airplane are.

The lift produced by the rotor aims straight upward for *collective pitch*, but with a device called the cyclic pitch control, the pilot can tilt the rotor blades to make the helicopter fly in a particular direction using *cyclic pitch* (**Figure 5.18**). Although most of the lift force still points upward, some of it now points to the front, back, left, or right, tilts the entire helicopter, and pushes it in that direction. Two disks, the upper and lower swash plates, transmit the intentions of the pilot in the cockpit to the rotor blades.

Figure 5.18 Helicopter's collective and cyclic pitch.

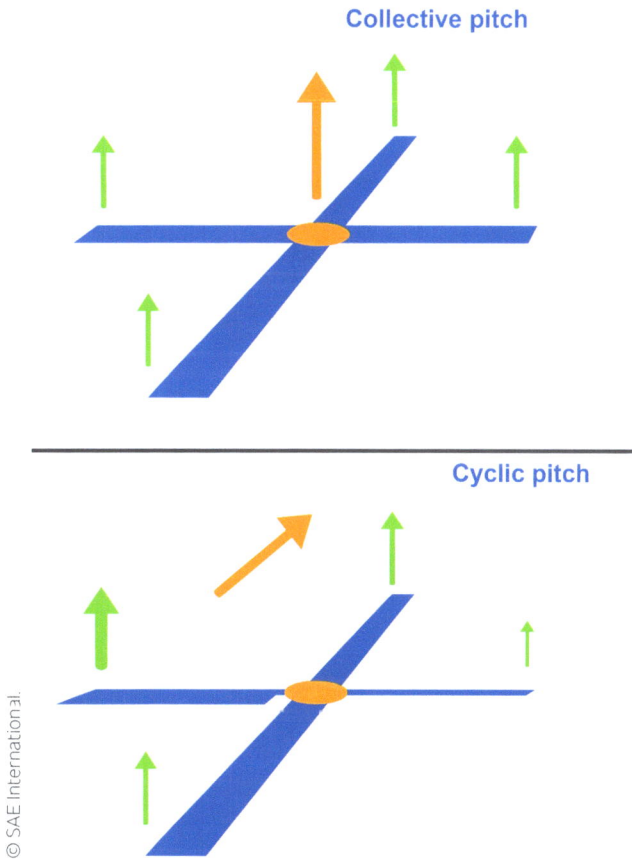

© SAE International.

The lower swash plate does not rotate but can tilt or move up and down. The upper swash plate spins with the rotors on ball bearings on top of the lower swash plate. When the pilot pushes the controls, the lower swash plate nudges the upper swash plate and the blades are tilted in turn by a system of control rods.

Everyone knows that a helicopter's rotors rotate, but did you know that they can also swivel back and forth as they rotate? That demands some impressively intricate machinery.

You can see what is going on by using your arms and your body. For the rotation, stand up with your arms stretched out horizontally and rotate your whole body slowly on the spot. For the swiveling, swivel your arms at the shoulders while you continue rotating.

You are now mimicking what a helicopter does with its blades, except that its blades swivel three to four times every second while they are rotating.

The main parts that achieve all of this are explained here:.

1. The blades are shaped like airfoils so they generate lift as they spin.

2. Each blade can swivel as it spins.

3. Vertical rods push the blades up and down, making them swivel as they rotate.

4. A central axle connected to the engine makes the entire blade assembly rotate.

5. The cap above the rotors is missile proof to protect against enemy attack.

6. Two turbo-shaft jet engines sit on either side of the rotors. If one engine fails, there should still be enough power from the other one to land the helicopter safely.

Why Do Helicopters Need a Tail Rotor?

According to the laws of motion, any force or action produces an equal force or reaction in the opposite direction. This means the torque (rotating force) produced by a rotor's blades tends to turn the fuselage (main body) in the opposite direction.

All helicopters have either a second propeller or another device to counteract the torque from the main blade. In most helicopters, a tail rotor balances the torque by pushing in the opposite direction to the main rotor.

Some helicopters have two rotors mounted on the same shaft, which counter-rotate to cancel the torque.

Miscellaneous Ways of Transportation

Jetman—an aerospace transportation vision—Yves Rossy had the idea and then built a wing-suit system comprising a backpack equipped with semi-rigid airplane-type carbon-fiber wings that spanned roughly 2.4 m (7.9 ft) (**Figure 5.19**). It was powered by four Jet-Cat P200 jet engines which he had made from large model aircraft engines fueled by kerosene. He was then referred to in the press by various monikers, such as the Airman, Jetman, Rocketman, and Fusionman.

Figure 5.19 Yves Rossy in his jet pack.

In 2008, Rossy made a flight over the European Alps, reaching a top descent speed of 304 km/h (189 mph) and an average speed of 200 km/h (120 mph). In November 2009, Rossy attempted flying across the Straits of Gibraltar, hoping to be the first person to fly between two continents using a jetpack. At about 1950 m (6500 ft) above Tangier in Morocco, he launched himself from a small plane and headed toward Atlanterra, Spain. The flight should have taken about a quarter of an hour, but because of strong winds and banks of clouds, he ditched into the sea. He was picked up 10 min later by his support helicopter, just 5 km from the Spanish coast, flown to a hospital in Jerez, and soon released unhurt. The Spanish Coast Guard retrieved the jet pack, which had its own parachute and a float.

A year later, Rossy had a new version of his jet-powered flight system ready to fly, and with it, he successfully performed two aerial loops before landing with the help of a parachute. This time, he launched himself from a hot air balloon piloted by Brian Jones at 2400 m (7900 ft).

Flight mechanics is the most applied part of aerospace engineering. Aircraft performance provides the foundation for making satellites an operational reality in orbit, aircraft as transport vehicles, and Jetman flying over the top of Mount Fuji. All of this occurs within the field of aerospace science. So now we will review the fundamentals before you learn more about this miracle that makes things work in the aerospace industry.

Conclusion

In this chapter, we have explored the basic physical laws that govern flight in air travel and atmospheric flight and how the two are different. In the next chapter, we will learn about the systems and methods for communicating with spacecraft.

Chapter 06

Communication with Spacecraft

Figure 6.1 Ariane-5 ECA launch of Herschel and Planck in May 2009.

© ESA/A. Chance.

Humankind is exploring the universe using all available means: powerful telescopes on Earth and in space to study stars, sending astronauts to the Moon, an ISS orbiting Earth, and even sending machines to other planets like Mars or Saturn.

This is great, but how do we talk to them when they are so distant from us? If we cannot talk to them, it would not matter what interesting things they might be doing because we would not know anything about them. So, at the end of the day, it does not matter if we send a person or a machine to do the job. If we cannot communicate with them, no science will be done as we would receive no interesting information about what they are seeing and measuring.

A lot of engineering is necessary to develop solutions to help us communicate with our spacecraft. In this chapter, we will talk about how we use electromagnetic waves to communicate with our machines and astronauts. We use electromagnetic waves to send information. Because space is "full" of vacuum, no sound can be sent back to Earth. Because electromagnetic waves travel very quickly, we can learn what is happening in places very distant from Earth in the quickest way possible.

Data Transfer Using Electromagnetic Waves

Electromagnetic waves are very strange. Both visible light (the colors you see with your eyes) and X-rays (used for making radiographies) are electromagnetic waves, but it gets even stranger. It turns out that mobile phones, television, and many other devices communicate by means of electromagnetic waves. We use the same type of waves to both see and communicate. How do electronic devices communicate using electromagnetic waves? How can we get information into them? By something called modulation.

Modulation of Electromagnetic Waves

Modulation is the changing of the features of electromagnetic waves so that they can be sensed by the receiver and that the information that was modulated into waves can be extracted. For instance, NASA's *Cassini* mission, which was recently orbiting Saturn, needed to modify electromagnetic waves to send us information about what it saw there on Saturn (**Figure 6.2**). Can you imagine if we had no means to talk to *Cassini*? We would not have been able to learn as much as we know now about Saturn.

Figure 6.2 *Cassini* spacecraft with ESA's Huygens probe.

Courtesy of NASA.

Cassini modified electromagnetic waves by using its transceiver, which is just a fancy name for the device we use to transmit and receive data by modulating and demodulating electromagnetic waves. From now on, we will refer to these electromagnetic waves as carriers as they carry information. After that, *Cassini* sent these electromagnetic waves very quickly back to Earth using its antenna. There, in a place we call the ground station, big antennas received a tiny signal, and the information *Cassini* modulated into its carrier was extracted. To talk to *Cassini*, we used the same big antenna and powerful amplifiers. An amplifier is a device that creates a powerful signal out of a weaker one. This is basically how we talk to spacecraft.

You might be wondering why we said we receive a tiny signal here on Earth. Fair question. Well, it turns out that once it leaves the spacecraft antenna, the electromagnetic wave distributes its power throughout space in an ever-expanding spherical shape. The further the waves go, the bigger that sphere becomes and the bigger the area the signal is distributed over (**Figure 6.3**).

Figure 6.3 Transmitted signal power expands in a spherical shape.

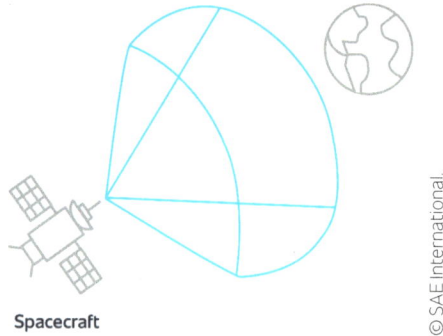

Spacecraft

© SAE International.

Our spacecraft is very distant from Earth, so the sphere is very big and the signal gets too tiny. When the wave loses its power, we say that the signals are attenuated. As a result, all our spacecraft need to be able to modify electromagnetic waves so they can send information to Earth and extract data from the signals coming from Earth. This is why all these machines are equipped with receivers (for taking the incoming signal and extracting information from it) and transmitters (for taking what they want to say and putting it into an electromagnetic wave).

They also need amplifiers and antennas. If you look at a picture of a satellite, you will always see some kind of antenna so it can talk with us on Earth. The top of *Cassini* in **Figure 6.4** has a white plate, for instance. That is its main antenna. When it comes to talking to distant machines, we should use parabolic antennas because they concentrate all the radiated energy into a small point in the sky. It is not very different from those antennas you can see on any house roof for watching satellite TV.

Figure 6.4 *Cassini* spacecraft in assembly.

Courtesy of NASA.

Ground Stations

You might be wondering what we use back on Earth to receive spacecraft signals. Special places called ground stations are in charge of receiving and dealing with our space data. As our machines are so distant from Earth, transmitted signals are so faint that it is very difficult to properly receive information. It is not only a matter of signal power, but time too. We are used to receiving instantaneous replies while using modern electronic gadgets, but when it comes to deep space spacecraft, it takes some time, sometimes even a day to receive a reply.

In order to communicate with our electronic explorers, it is necessary to apply a wide variety of complicated telecommunications techniques. It is interesting to note that many of these deep space techniques are applied to modern everyday systems. The best example of this is with current mobile phones: LTE (4G) technology uses physical principles derived from arraying techniques, which have been studied in the aerospace sector to overcome the fact that big receiving power is necessary for receiving weak signals from spacecraft. We will talk more about this later.

If you have ever come across a ground station, you will have seen big dish antennas pointing to the sky. The farther the spacecraft is, the bigger the antennas need to be. For instance, both NASA and European Space Agency (ESA) use ground stations networks to communicate with all these machines that are so distant from Earth. They use 34- and 70-m-diameter antennas to receive and transmit information—the heights of 12- and 23-story buildings. These ground stations are placed around 120° of longitude from each other in order to operate 24/7 without losing a single bit of data (**Figure 6.5**). If you take a look at the globe of the Earth in **Figure 6.6**, you will see why.

Figure 6.5 Redu Ground Station.

Figure 6.6 Three points are necessary to cover 360° as Earth rotates.

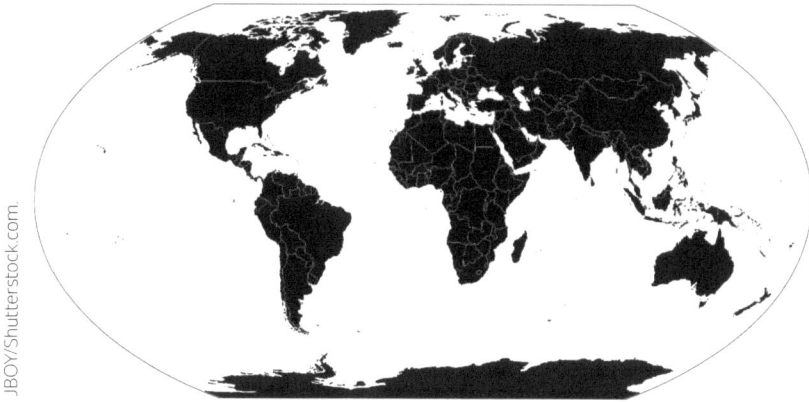

Depending on what type of spacecraft we want to track, we place our ground stations at different locations. For instance, if we want to track satellites orbiting Earth with polar orbits, which is widely used for weather and defense satellites, we will place them as close as we can to the Earth's poles. This will ensure the maximum possible number of satellites pass over the ground station. This is why you can find them in places like Svalbard, Norway.

In **Figure 6.7**, you can see a ground station in Svalbard. The white bubbles are called radomes. They cover ground station antennas and protect them from bad weather like snow, wind, etc. Nevertheless, we need more than huge antennas and many ground stations if we want to communicate with other planets.

Math is mandatory for communicating with other planets. We will discuss further some nice tricks we can use so we can hear our spacecraft. At the end of the day, talking to a distant spacecraft is pretty much like talking to some friends in a noisy place: you have to use your imagination to find out ways to be able to talk to them properly (**Figure 6.8**).

Figure 6.7 Svalbard Ground Station under the Aurora Borealis.

© ESA.

Figure 6.8 Polar orbit.

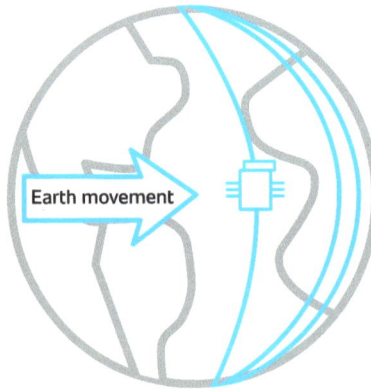

Earth movement

© SAE International.

One thing we usually do is cup our hands behind our ears to hear better. By doing this, we are somehow amplifying the incoming signal. This is the main point of having huge antennas: they are like having huge ears for listening to the sky. It is important to note that sound and electromagnetic waves are not the same thing at all, but big ears are a nice way of visualizing the purpose of these antennas, though.

Sound waves are mechanical waves, so they need a medium for propagating through space. The same thing does not apply to electromagnetic waves, so they can be used for telecommunication in space. By looking into how these waves behave, we can even learn things about other planets, as you will see later on in this chapter.

Let us go back to our example of talking to some friends in a crowded place. You could try to shout to be heard, which is why very powerful amplifiers are used at this kind of ground station. NASA also has spacecraft outside the solar system, like *Voyagers 1* and *2*, which left Earth in the 1970s containing a gold disk of information about us, designed by Carl Sagan.[1]

Now things start to get tricky! Another way to help your friends communicate is to use a made-up code, for instance repeating each phrase twice. This will increase the probability of being heard. If your friends do not hear you the first time you say something, they might the second time. They could even get one part of the sentence in the first try and the other part in the second. By doing this, you are adding what is called redundancy to the message.

Modern telecommunication systems use redundancy to enhance communication. Special codes called forward error correction codes are designed to add redundancy in a controlled manner to messages. It does increase the amount of information that we send through our channel, meaning it doubles if we are sending the message twice as in this simplified model, but on the other hand, it will increase the probability of successful communication. Success is measured by

---

1 If you do not know about Carl Sagan and the Cosmos series, now is the time to Google it.

engineers using a test called a bit error rate (BER), which allows us to measure the probability of receiving a bit of digital information wrongly. By applying forward error correction codes, we will get a better BER ratio.

Antennas

There are many types of antennas. If we want to know how one particular antenna radiates, we need to look at its radiation pattern. Radiation pattern calculation is quite complicated, but we will look at some radiation pattern types. Depending on the application, we need to radiate to a single point or to all points in the sky. For example, if we want to use an antenna in a Wi-Fi repeater, we will need one that radiates to all points around it. There are two ways of increasing the signal gain of the antenna being used. We can use one big antenna and/or synthesize the signals from an array or group of separated antennas.

Big Cassegrain Antennas

This Cassegrain antenna is an enormous structure designed to concentrate electromagnetic waves into a single point (**Figure 6.9**). Its biggest part is the main reflector, which captures electromagnetic waves, so the bigger this part is, the greater the gain will be. We define gain as the ability to amplify incoming signals. After bouncing off the main reflector surface, signals go to the sub-reflector, which is the metallic structure above the antenna, fixed to the ground by the four metallic legs. Signals are then sent to the center of the main reflector, where they are typically captured and passed to electronic systems.

Figure 6.9 Close view of a big Cassegrain antenna.

© ESA.

Since radiation pattern calculation is complicated, we cannot go into much detail about it here, but we can at least use a valid approximation to talk about the relationship between how big a dish is and the gain that an antenna has:

$$G \propto \left(\pi * d / \lambda \right)^2$$

where

d is our antenna diameter

π (pi) is the irrational number

λ is our signal wavelength

From the equation, we see that the bigger the antenna is (d is in the numerator) and the higher the frequency is (higher frequencies have a smaller wavelength[2]), the more the gain will be.

---

2 There is a well-known inversely proportional relationship between frequency and wavelength. As wavelength is in the denominator, a small wavelength value implies higher gain. Higher frequency therefore implies higher gain.

Therefore, according to this equation, we know that we want big antennas and higher frequencies. An antenna that is 70 m in diameter can amplify a maximum of 551×10^5 times[3] the incoming signal. (Please note that there are technical issues that make big antennas operating at high frequencies complicated, but we are not dealing with them here.) An antenna that is 34 m in diameter can amplify a maximum of 129×10^6 times the incoming signal at a frequency of 32 GHz. As this equation does not take into account many issues that occur for real antennas, these are maximum values that will not be met in reality.

Wave Features

When it comes to signals, everything has to do with waves. Because of this, concepts like frequency, amplitude, and especially phase need to be well understood to really master what is going on in a parabolic antenna. In order to properly address this matter, we will do a quick review of waves. We will start with a classical wave (as shown in **Figure 6.10**) and will then talk about wave features.

Figure 6.10 Sinusoidal wave.

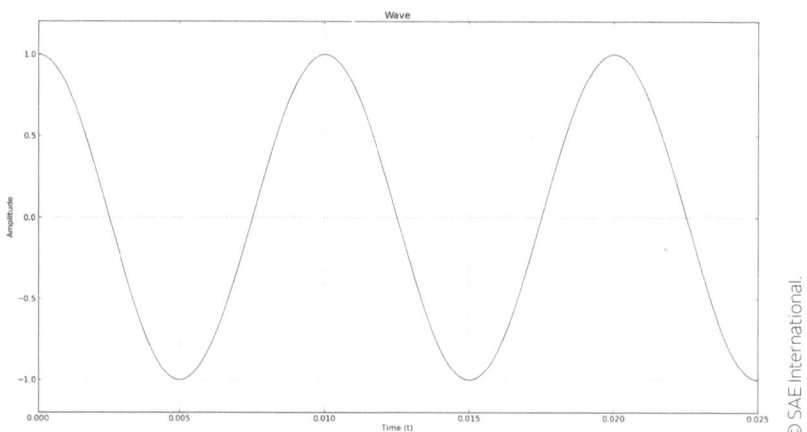

© SAE International.

---

3 In radio frequency engineering, these amplifying factors are not expressed this way. A special unit called a decibel is used. Then, theoretically, a diameter of 70 m at 32 GHz will give us 87 dBi, and one of 34 m at 32 GHz will give us 81 dBi.

Amplitude: This is the maximum value the wave reaches on the vertical, i.e., y-axis. In our example, at point 0.0, $A_0 = 1$, amplitude is related to wave power, so the higher the amplitude, the more powerful the wave is.

It turns out we can modify wave characteristics such as wave amplitude to send information. This is called *amplitude modulation* (*AM*), and it is the simplest way to modulate the wave. The distance (in space, not time) between two consecutive peaks of a wave is called the wavelength, denoted by the Greek letter lambda, λ.

In **Figure 6.11**, the wave amplitude is formed by the former amplitude of the faster blue wave *times our information* (which this time, as a matter of fact, is another wave, the red wave). Therefore, to use AM, we just have to multiply the information and the wave. The faster blue wave is the carrier, and the other red wave that modifies the carrier amplitude as time goes on is the *envelope*. But amplitude is not the only feature we can modify to transmit information by means of electromagnetic waves.

Figure 6.11 Amplitude-modulated wave.

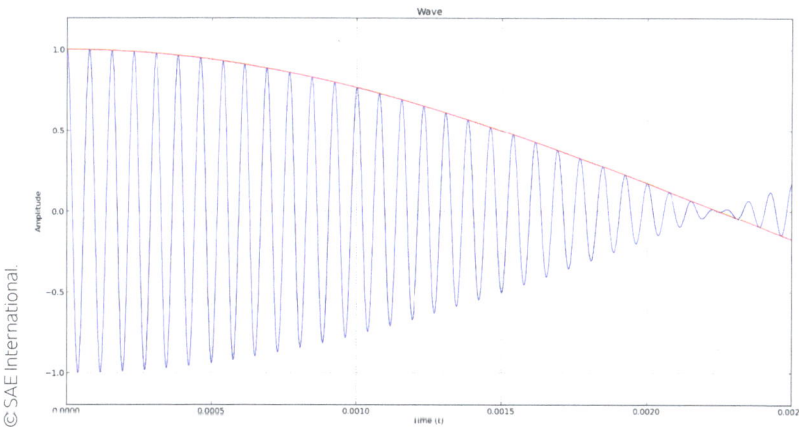

Phase: This is the oscillation state at any given point, both space (x[4]) and time (t). The same values will occur many times in space and time, as simple waves are periodic. λ and f will tell us how long it will take for

---

4 You may have noticed our figures only depend on time t and not space x. This is because in order to make examples easier to understand, we are just using 2D figures. You can assume we are looking at a single point on the x-axis, for instance $x = 0$.

a value to be repeated in both time (f) and space (λ). Phase is very important when it comes to combining signals. Let me show you why.

By observing the wave in **Figure 6.11**, we will see what happens if we change its initial phase from 0 to π in **Figure 6.12**. The green wave can be seen to be moving the blue wave a little bit to the right or to the left in the graph. **Figure 6.12** has the original blue wave and the same wave with an initial phase π greater in green. As you can see, their values are opposed.

Figure 6.12 The importance of phase.

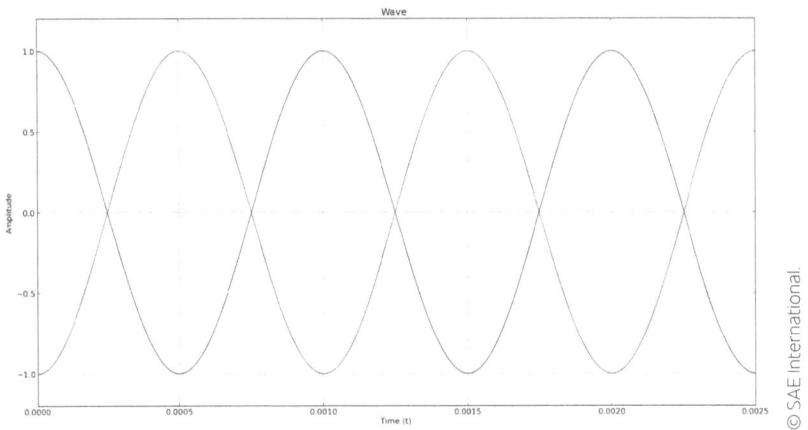

If we tried to combine them by adding them together, we would obtain a zero value for all given times,[5] resulting in a so-called *destructive combination*. We could measure a phase difference at any particular point by comparing the wave oscillating state at that point to a given reference. In **Figure 6.12**, you can see both waves are not equal. They have the same frequency and amplitude, but a different phase. If the phase was the same, then adding them together would give us a wave with a greater amplitude, which would be a powerful wave, i.e., a powerful signal. This is called *constructive combination*. We need to pay great attention to phase, as it is important for combining signals. You will see this in the section Arraying Antenna.

---

5 And space.

Modifying wave phase is also a good method for modulating a carrier and sending information. In fact this is widely used in the aerospace industry and is called *phase modulation* (PM). In the first graph in **Figure 6.13**, you can see a binary signal (1 and 0 values) represented by a red line. We superimposed this value over a normal cosine wave and then merged them. You can see the result in the lower graph. Note that every time the red line rises or lowers (representing a "1" or a "0," respectively), the phase of the carrier changes in the result.

Figure 6.13 Phase modulation.

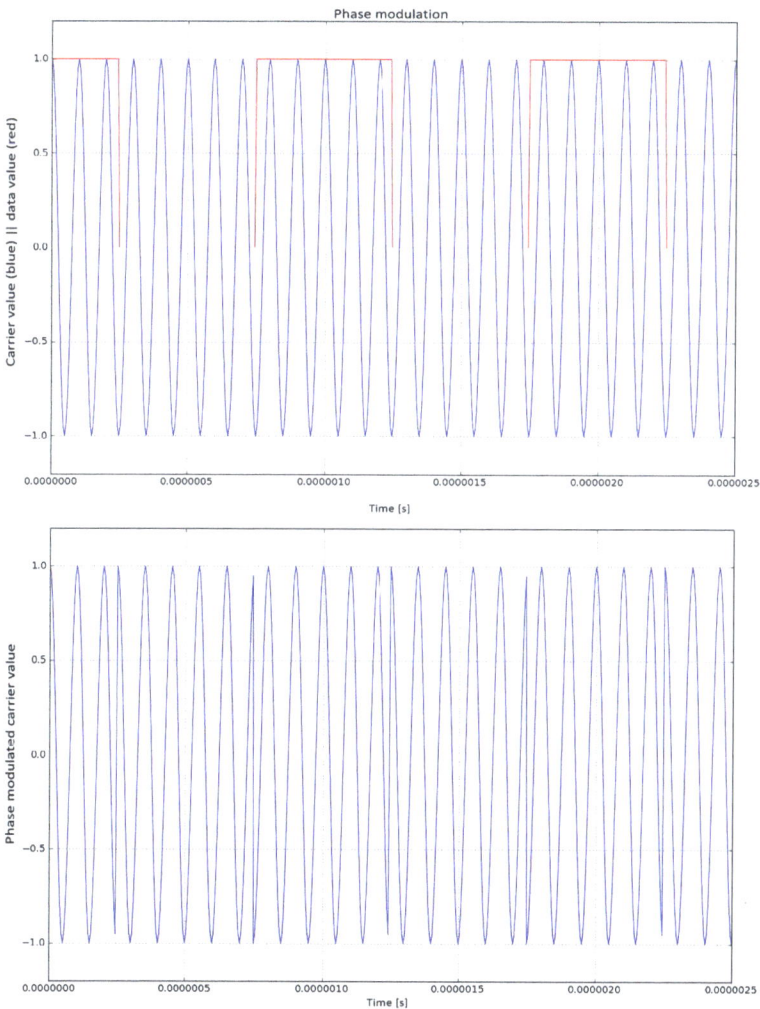

Frequency: This is also a very important characteristic of a wave (**Figure 6.14**). It lets us know how many cycles per second a wave is performing. It is measured in hertz [Hz]. We can also modify carrier frequency to encode information into an electromagnetic wave. This is called *frequency modulation* (FM). There is an inversely proportional relationship between frequency and wavelength. All these wave features can be modified to transmit information. The techniques for this are called modulation techniques.

Figure 6.14 Sinusoidal waves with two frequencies.

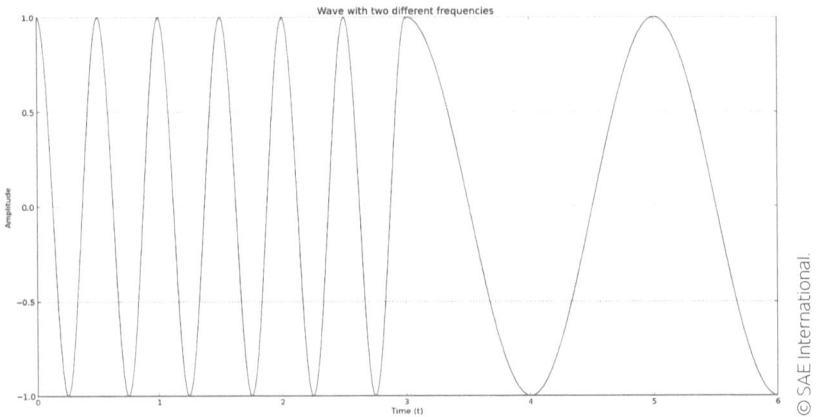

Other Things We Learn from Wave Features

It turns out that we can learn other things based on these three wave characteristics (amplitude, phase, and frequency). An electromagnetic wave undergoes some modifications of its characteristics depending on the medium it is going through and the interfaces between those mediums.

Because of this, we can even learn about another planet's atmosphere by sending a signal from our spacecraft to Earth, if the signal travels through its atmosphere. This is called the *Radio Occultation Technique*. In order to do this, it is necessary to first extract the Earth's atmospheric influence out of the signal so that we can distinguish between modifications introduced by its atmosphere and those that came from far beyond the Earth's protective atmosphere.

Parabolic Antenna Shape

Now that we have a deeper knowledge about waves, we will return to our big Cassegrain antenna and analyze its shape. It is called a parabolic antenna because of the shape of the main reflector. It is designed with this particular shape to ensure a constructive combination can occur at a particular point on the antenna. Let us take a closer look at the antenna's shape to truly understand what is going on.

Mathematically, Q_1P_1F, Q_2P_2F, and Q_3P_3F are of the same length, and this applies regardless of which P point you decide to use as long as P is on the parabolic-shaped surface. This is important because this means all the paths our signal could travel along from the plane of Q to point F, that pass through the parabolic surface, are of the same length. So, if the waves are at the same oscillation state at plane Q, then their phase will be the same at point F too. Therefore, we can collect many waves present at plane Q and concentrate them at point F.

As the oscillation state will be the same for all of them at F, a constructive combination will take place, and as we have seen already, this means we will obtain a powerful signal. The higher the number of waves that are combined at F, the stronger the signal will be. If our antenna is big, more waves present at plane Q will be steered to F and properly combined. We therefore end up with the known conclusion: the bigger the antenna is, the greater its amplifying factor.

Arraying Antenna

As we have seen previously, it is advantageous to concentrate signals and combine them with the same phase. And the bigger the area we use to collect these signals (i.e., the bigger the antenna), the bigger the amplifying factor, as long as we combine them in phase.

Then, Why Would We Use Just One Antenna?

If we use more collecting surfaces and then combine the collected signals with the same phase, we will mathematically create a bigger antenna. It is possible to use two big and separated antennas to record a signal coming from a far point and then combine them with the proper phase. To do this, we have to generate the same effect a parabolic surface creates in waves that bounce on its main reflector.

We would also have to introduce a particular extra delay, as combining signals from two distant antennas involves another issue. It turns out when it comes to signals from a point far distant from Earth and antennas are distant from each other, the wave front (this is the real name of plane Q in our explanation about the parabolic shape for **Figure 6.15**) will not reach both antennas at the same time. This can be seen in **Figure 6.16**.

Figure 6.15 Parabolic shape reflection.

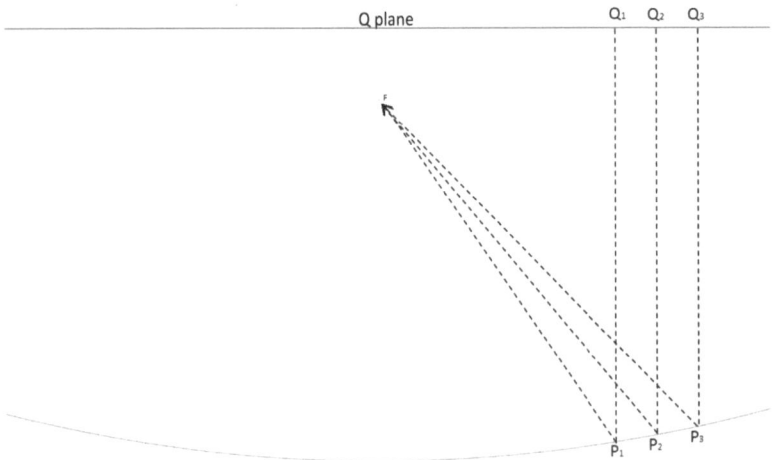

© SAE International.

Figure 6.16 Geometric delay between two ground stations receiving the same spacecraft signal.

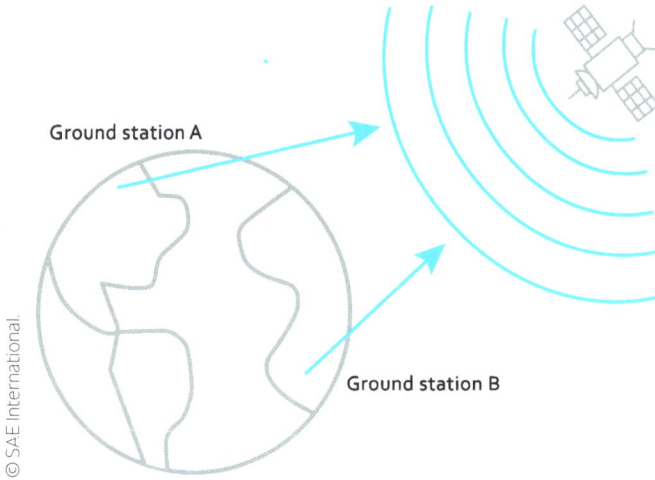

This is an issue. Because there is a delay in the system between ground stations A and B, if we try to combine the signals recorded at both antennas for a given time without taking the delay into account, we will not obtain a constructive combination. We will not be sensing the same oscillation state and therefore not obtain the desired constructive combination. We would have to record signals during a period of time and then try to introduce a delay into one set of them, to minimize relative phase[6] between both of them. By doing this, we will maximize wave amplitude in the resulting combination.

What we are trying to do is correct the geometrical delay the signal is undergoing. In order to properly do this, the influence of the atmosphere above both antennas has to be taken into account. An interesting consequence of this fact is that in order to obtain the proper geometrical delay, meaning in order to know the path difference between the two antennas as seen by the wave, it is necessary to know the horizontal distance between both antennas.[7]

---

6 Phase difference between both waves.
7 This is a trigonometric issue related to the signal source point in sky and the horizontal distance between both antennas.

This horizontal distance has to be accurate in terms of wavelengths, which uses distances beneath the centimeter scale. This combination also allows us to learn about the movements of the Earth's tectonic plates: by the trial-and-error method, we can introduce different delays to the signal until we obtain a maximum value in the combination. This way, we will know the distance between antennas. If we monitor the evolution of these distances over time, we will be measuring the movement between ground stations or between two sites on Earth's surface. This physical principle is widely used in geodetics (the science of measuring the Earth).

Three Data Types in Spacecraft Telecommunication

A ground station performs three main tasks when it comes to spacecraft on missions: tracking, telemetry, and command, known as TTC.

Tracking
The ground station follows a spacecraft as it goes through the sky above our heads. Meanwhile, it receives some interesting data about the spacecraft's movement and position. This data is very important and will be used to decide orbital maneuvers.

It is necessary to avoid any possible crash with planets, moons or other spacecraft while keeping our spacecraft where it is supposed to be. In fact, watching satellite TV is only possible because a group of people worked very hard to make sure the satellite is positioned where it is supposed to be so you could point your antenna to a well-known point in the sky and watch your favorite TV shows. This applies even for geostationary satellites which are supposed to be always at a fixed point in the sky. Orbital station-keeping maneuvers are necessary to ensure satellites remain where they are supposed to be.

Doppler: It is a physical principle widely used in modern devices. For example, doppler is used with radars to learn the radial speed (how quickly the target is getting closer to us or further away from us). Radial speed slightly modifies the way we perceive the received frequency coming from the spacecraft. If the spacecraft is moving closer to us, we will perceive a slightly higher frequency, and if it is moving away from us, we will perceive a slightly lower frequency.

Have you ever wondered how astronomers know the universe is expanding? Here is the answer: the Doppler effect! Radiation coming from stars moving away from us has a slightly lower frequency than it should have. When it comes to visible light from those stars, we see non-red-colored stars in red (or with an unusually higher red content), as those frequencies are lower than they should be. This is called *redshift*. *Blueshift* happens when an object is moving closer to us.

Sequential Ranging: We know how fast our spacecraft is moving in the radial direction. But how far away is it? This is a very interesting question, and we have another tool called ranging that will help us to know.

Now, imagine you are in a dark cavern. You do not know where the walls are and you are unable to see them. What could you do in order to know how far you are from the wall? If you had a baseball, then you could throw it at the wall (let us assume you know what direction you should throw it). Your ball hits the wall quickly and then comes back to you. If you had a good clock to measure how long it took for the ball to get back to you and you knew the ball speed (that is the tricky part), then you would be able to work out the distance between the wall and yourself. Depending on how accurately you can measure time with your watch, you will be able to measure the distance to the wall very precisely. Let t_b be the time it takes the ball to get to the wall and back. Let v be the ball speed (let us also assume it is constant the whole time even when it is bounced on the wall). Then we know that

$$d_b = (v_b x t_b).$$

If you want to know how far you are from that dark cavern wall, you have to take into account that t_b is two times the distance to the wall, as your ball traveled to the wall and back again. So if d_w is the distance to the wall, then

$$d_w = d_b / 2.$$

For spacecraft, we know how fast light travels in vacuum: it is a well-known physical constant, C_0.

Now let us get back to satellites. A special signal, called a ranging signal, is sent with data. Spacecraft need some time (a very small amount of time) to get that signal and send it back to Earth. So we measure how long it takes for the signal to get to the spacecraft and how long the spacecraft needs to process this signal and send it back to Earth. As it was for you standing in the middle of a dark cavern, measuring time accurately is very important.

Because of this, a special type of clock called an atomic clock is needed. It allows us to accurately know when we send our ball (the ranging signal) to our spacecraft and when we receive it. What we will measure is in fact phase differences. (If you are not sure what a phase in a carrier signal is, see section Wave Features in this chapter.) In order to avoid inaccuracies resulting from the fact that you do not know how many entire cycles your signal has undergone before getting back to you, different frequencies are used. The larger the number of frequencies, the smaller the inaccuracy.

Delta DOR: We are almost there now. We know how fast our spacecraft is moving, and how far it is. But where in the celestial sphere is it? Should we point antennas to the south … east …? Well, you will have had a basic idea from the very beginning because if for instance you are tracking something orbiting Mars, your spacecraft will be positioned where Mars is positioned in the sky. That is fine, but if you want something more accurate than that, if you are receiving your maximum signal[8] you can be very sure about where in the sky your spacecraft is (depending on your antenna radiation pattern). However, we will need something even more accurate than that in order to land safely on the Martian surface. How can we decrease that error? By using something called Delta DOR, which is based on arraying. We will perform measurements using two ground stations at the same time, and then by combining the signals properly, we will know very precisely from what point in the sky the signals are coming from.

---

8 Remember the radiation patterns.

Telemetry

All spacecraft instrument data and spacecraft status are sent back to Earth as binary data. Raw data that comes out of spacecraft devices (cameras, sensors, radars, etc.) is sent to Earth and is known as telemetry data. After processing this data, the science we read in newspapers is generated.

Command

Every time we want to order the spacecraft to do something, we will transmit information to it. Every time we want to tell the spacecraft to do something to turn on a thruster, activate some instrument, or maybe perform a software update, we will do it by sending data to it. This is widely known as command.

Conclusion

In this chapter, you learned about the methods that we use to communicate with spacecraft, as well as some of the physics underlying those techniques.

In the next chapter, we will be shifting our focus to ways that you can get more involved in the aerospace industry and boost your career.

Chapter 07

Career Overview

Figure 7.1 Final assembly line at Airbus' Hamburg plant.

In the Aerospace Industry

Jens Strahmann, the former head of the Airbus High Lift Department and the current vice president of PACAVI Aircraft Conversion, advises what a career in aerospace engineering would look like, what experience in this industry would bring, and how being open-minded helps us move along with the tremendous technology of the aerospace engineering world.

What Is It Like to Be a Scientist/Engineer in the Aerospace Industry

We will look at this question from a number of different angles.

1. Experience: what does it mean?
2. Projects come and go.
3. Synergies: using new techniques in an airworthy environment.
4. Being open-minded and critical.
5. What is doable and what will remain as an idea only?
6. Not invented here.
7. Young professionals and old managers.
8. Small, medium, and large companies.
9. Team and individual.
10. Timing for new ideas.
11. Innovation cycle: aeronautics, astronautics, and automotive.
12. How do I start?

Experience

Experience means learning from the bad and the good, failure or success. The experience of failure you will never forget and is character forming. Experience from success grows your motivation and provides a positive attitude. Feedback is the tool for improvement. Even negative feedback can be good feedback.

The right level of experience is key:

- Too much will reduce your decision-making risk.
- Too little will result in trivial mistakes.

Suggestion: A senior mentor who helps eliminate trivial mistakes but encourages decision-making ability is important for a young professional's career.

Projects Come and Go

New projects are the fuel for energizing your mind-motor. Every scientist is interested in challenging new projects, but be aware that you are dealing with projects of the right size. In aeronautics or astronautics, the realization cycle can be especially long.

Young scientists have started their careers on huge projects in big companies, and the project has come to fruition 25 years later—or worse, was canceled after 10 to 15 years, ending in complete failure. Even if they gained a vast amount of experience, in the end, it was for nothing and could mentally drag them into a big hole. Starting with projects that have time schedules of several months may be better. Starting in small-/medium-sized innovative organizations provides you with fast and direct feedback, which is essential for your future personal development. Really new and innovative ideas are coming from small/medium enterprises anyway.

Synergies: Using New Techniques in an Airworthy Environment

By starting in small companies, you will acquire a lot of innovative, sometimes disruptive, knowledge that is key for your future career. This knowledge could be essential for your next profession. Even, at a first glance, the acquired skills have nothing to do with the new project very often they do. It might be the tools used, the processes, the network of people and companies, or the mindset. In fact, using disruptive knowledge from an earlier project paired with a new subject could raise synergies nobody has envisioned.

Flight control for nearly 100 years was dominated by analog techniques, but by adding the up-and-coming digital processors, reliable data storage, and specific program languages, it has now become possible to control aircraft using fly-by-wire (i.e., totally electronic) techniques. This used to be a disruptive brand-new approach. Today, nobody would even argue about the feasibility of digital aircraft control, and soon, passenger aircraft will be flown unmanned using a combination of next-generation network technologies, fly-by-wire, GPS, and satellite communications.

Being Open-Minded and Critical

Being a team player does not mean you do not challenge obviously wrong decisions. The secret is how you challenge them. It is very important to discuss the issue in a very respectful way, even if you are young and you are 100% sure to be right (what else!). Being open-minded, asking, and trying to understand are factors in being an exciting person. You are leading by asking and showing social competence if you retain the talent for being open-minded and cooperative.

What Is Doable and What Will Remain an Idea Only?

Pretty often great ideas get stuck in the company's gear box. The traction is somehow too weak. The decision process takes forever. Momentum is soon gone. Innovations brought to the market, like the iPod, iPad, and iPhone, have been great successes, but the company that won the race was not the one that had dominated the market for several years.

Compare Nokia and Apple. Both companies had different starting points for approaching their smartphone. Apple had a convincing, innovative but risky approach. Nokia chose the safe way (never touch a running system) and lost completely, even though the prerequisites were clearly on their side. The consequences for Nokia were dramatic: the wrong path led them to insignificance.

The innovation cycle for smartphones is very short, and once you lose track, even money cannot rescue your company. This was "unpleasant" for the upper management but dramatic for all their employees. Senior technical engineers at Nokia knew that for sure, but they could not convince their managers to accept their arguments.

Despite the fact that in the past technical directors guided companies, today it is financiers, lawyers, and economists who are in the driver's seats. Money, no risk, "let's take the safe way," and a lack of real research were the guiding principles. Only startups with disruptive ideas can be the real innovators. The results of this phenomenon can be seen today. The only chance the big companies have is to buy the startups and consume their products.

This is happening even in the aerospace industry. The modern light-weight battery technology inspired by the satellite industry was developed by computer, cellphone, and small car manufacturers. Modern innovative LED lighting came from consumer electric appliances, not from aircraft cabin lighting companies or even original equipment manufacturers (OEMs). Three-dimensional-printed parts are used all over the industry, and finally (half-heartedly), the aircraft industry is adopting the technique but is hiding its more innovative usage behind airworthiness regulations.

Not Invented Here

Especially in large enterprises with multiple sites clustered across several locations and/or countries, it becomes even more challenging to share your inventions or ideas for solving a problem if you are not working at the central location. The journey could be very long and exhaustive. A solution might be a mentor, a sponsor, a network of open-minded colleagues, or better a gray eminence. None of these supporters come for free. Meetings, training, or coaching will help to form a group of supporting staff. This will help to overcome the phrase: "Not invented here in the center." Even if the management will never admit that the company works centralized, you can be 100% sure it is!

Young Professionals and Old Managers

What is the difference between a young professional and an old (experienced) senior manager? It is how they approach a task or a problem. In a long-lasting career, an experienced manager has developed a method for solution finding. It is a method formed by failure, success, and the decision-making process and has cost the company a lot of money. Recently, tools have been created to support systematic failure analysis, knowledge transfer, and retention of knowledge. All these processes could help minimize the legacy process of trial and error.

The idea behind them is to shorten the experience cycle. Good intentions are sometimes poorly realized. Much better would be the pairing of a senior manager with young talent for a dedicated transition time and follow-up period. This would need strategic personnel planning and future market orientation, which might be difficult in an industry environment dominated by short-term and/or extremely short profit and loss cycles. After all, what kind of strategy can you form if you do not know if the company will still exist in the next three months?

Unfortunately, industries and OEMs with medium/long innovation cycles do not behave differently—not even when it is really necessary.

Small, Medium, and Large Companies

There are different advantages and disadvantages to working in small, medium, or large companies.

For mainstream companies (medium or large organizations), the pros include the following:

- An uncluttered organization.
- An established hierarchy.
- Existing processes.
- Existing tools.
- Educated staff.

For a specific organization, the pros include the following:

- A dedicated group with special knowledge.
- Smaller/more effective units.
- Better change processes (i.e., better reactivity).
- More flexible units.
- Dedicated documentation.
- More effective configuration control.
- Independence from "elephant" processes.
- Prototype character visibility.

There are pros and cons for both organizational types. A lesson learned in big organizations with the complicated interface between development, production, and flight testing has not proven to provide the most effective type of certification within the shortest time.

A better organization therefore might be a dedicated, clearly arranged center including specific production and specific development with a dedicated, responsible test center. This would ensure minimal interface issues and minimal interferences with production and would not disturb the mainstream manufacturing process. This might be the leanest approach.

Team and Individual

Employment will usually start in a team approach. Learning from others is smart, but doing things the same way as others is not smart enough. It is important to find your own way of working and being a valuable add-on to the team effort. This is not limited to expertise or knowledge: personality and social competence are of equal importance. Inspiring the team with your specific personality is sometimes even more important.

For example, if your view on a problem is optimistic and solution-oriented rather than pessimistic, it makes your mind open to nonstandard solutions. As someone wisely said, "The difference between an optimist and a pessimist is that a pessimist is an optimist with too much experience."

The real assets of successful enterprises are their personnel, but on the other hand its biggest problems come from those personnel as well. That sounds paradoxical initially, but only an effective balance between both will be successful.

Real team spirit is the basis of an enterprise, but do not forget that a team consists of individuals who are different and want to make a career. Nearly all employment starts in a team in which people take responsibilities, and only real team players will be successful.

Timing for New Ideas

Having a great idea does not automatically mean you are successful. The timing for realizing ideas has to be right too. Very often ideas are brilliant, but they do not match the company's environment. What does that mean? In huge operations, idea spaces, idea processes, idea brokers, idea integrators, and so on are formalized in order to canalize the ideas for evaluation purposes.

Senior experts judge and filter new ideas and pick up or dismiss them. Sometimes good or bad coincidence can judge your concept, which can be frustrating because those experts do not have the vision of the originator. The time required for presenting ideas is sometimes limited, and some aspects do not get recognized as a result. What to do?

- Find a sponsor for your idea.
- Use your network and enlarge it.
- Present your concept/idea on a level so others can understand. Do not be too technical—make it easy to understand and simple to repeat to others.
- Be engaged and show confidence, and show your eyes sparkling with joy.

If these suggestions do not work, you have only two options:

- Drop the idea—or if you are convinced about your idea.
- Put it into a broader space outside of your direct sphere of influence. Find people and organizations with the same spirit, and make it your own business.

One thing is not allowed: being frustrated/destructive and belonging to the other 60% of the employees who have already internally quit their jobs. That is not good for the economy and your company, and most importantly, not good for you!

Innovation Cycle: Aeronautics, Astronautics, and Automotive

An innovation cycle defines an idea or an invention that has been realized economically. In the 1950s, 1960s, and up to the 1980s, astronautics was the leading technology driver and innovations appeared everywhere: moon landing, satellites, shuttle, and space station. Technologies stemmed from those huge programs and influenced navigation, communication, flight guidance and control, computer technologies, and digitalization.

Digital flight control was one of the major innovations that inspired aeronautics. The reproduction of systems, lightweight computers, digital navigation, reliable data storage, and digital communication made it possible to establish fly-by-wire in commercial transport aircraft. Today, all new developments for commercial aircraft are fly-by-wire, instead of the old analog flight control. Astronautical technologies have advanced by decades compared with automotive technologies.

That has changed in the last few years. Electric-driven and controlled cars (hybrid is only an intermediate step) will dominate the market in the near future. The force of competing companies, scientists, engineers, and money involved is enormous and fuels shorter innovation cycles. Efforts to research energy storage, 3D printing, and autonomous driving are relevant for the future of any car manufacturer.

Even German car manufacturers are investing a lot of money in these developments as they know full well that their current leading position in car manufacturing worldwide is in danger. The innovation cycle started by Tesla is heating up and has started a revolution rather than an evolution. The car will not only be a transport but also communications and living center in the future. If you miss the trend or an innovation, you could be soon out of the market no matter how big your company is. As a matter of fact, astronautics/aeronautics will use innovations and new developments stemming from automotive in the next decade, unless new horizons come from deep space exploration or manned missions to other planets.

How Do I Start?

This is always a good question. Often, the university or the company has been named the most important factor. That might still be the case, but only for exposed non-open positions. Today, the most successful and shortest way is to take the initiative yourself! If you have the chance to talk to engineers/managers during a conference, fair, exhibition, or talk, use it to let others comment on you. Be engaged, a bit devoted, friendly, optimistic, and interested in everything. If a contact is started that way, hang in there and commit to an internship, hands-on training, master's thesis, or other employment. Never ask for money or a vacation. Money will come, and vacations will happen.

It would be helpful to know what you want. That sounds easy, but in most cases it is not. For example, a young pilot in education had to wait for his job because of a waiting list. He could have waited forever, but after seeing a sign on the wall of a building saying *Flight Test*, he invited himself to the company and managed to talk to the senior flight test engineer. He explained where he was coming from and that he wanted to work for the company doing no matter what—because he wanted to fly missions and everything that is close to that. After a few appointments, he managed to work for the company and had the experience of his lifetime. Being heavily involved in development testing for an aircraft prototype, only a few years later, he was the lead technical pilot for a big airline.

If you know what you want and demonstrate engagement, you will gain the commitment of someone, who will act as a mentor. Somebody promoting you is a factor more effective than you promoting yourself.

Make the difference—good luck!

Chapter 08

Startups: Revolutionizing Aerospace Engineering

Figure 8.1 SpaceX achieves its first-ever "Chopsticks" landing.

Photo Credit: Steve Jurvetson. Licensed under CC BY 2.0, https://creativecommons.org/licenses/by/2.0/.

Introduction: Why Startups Are Redefining Aerospace

Aerospace engineering has always been at the frontier of human innovation, but in recent years, startups have disrupted what was once a domain dominated by governments and massive corporations. Companies like SpaceX, founded by Elon Musk, have shown the world that bold ideas, determination, and a startup mindset can reshape the future. SpaceX did not just redefine rocket science; it redefined the business model of aerospace, making reusability, speed, and scalability the new norm.

For me, the journey into aerospace was deeply personal. Growing up, I was fascinated by the intersection of engineering and problem-solving. Founding Plasmos Inc. was born out of a dream to create systems that would make space more accessible and impactful. At Plasmos, we tackled one of the most challenging problems in the industry: building a reusable third-stage rocket—the SpaceTruck. Designing it to deliver payloads across orbits while ensuring cost efficiency taught me the importance of agility, persistence, and vision. Startups, like Plasmos and others, prove that even the most ambitious ideas can take off when fueled by innovation and grit.

This chapter will guide aspiring aerospace entrepreneurs through the key areas that startups can innovate, from revolutionizing manufacturing to leveraging AI. Whether you are a student dreaming of building the next SpaceX or a professional looking to disrupt traditional models, this is your call to action. The sky is no longer the limit—it is just the beginning.

Manufacturing Revolution: From Earth to Orbit

The aerospace industry is in the midst of a manufacturing revolution, and at the heart of this transformation is the pioneering work of SpaceX. By vertically integrating production, SpaceX redefined efficiency, allowing for rapid iteration and cost reduction. The Falcon 9 rocket became the first reusable rocket to land and relaunch successfully, changing how the industry views sustainability and profitability. The Starship program, built using advanced stainless steel and robotic welding, pushes these boundaries even further, showcasing how startups can lead in rethinking materials and processes.

Additive Manufacturing: The Game Changer on Earth

Additive manufacturing, commonly known as 3D printing, has become a cornerstone of modern aerospace innovation. SpaceX integrates 3D printing to produce complex engine parts that are lighter, stronger, and faster to produce. Relativity Space, a startup founded with the vision of automating rocket production, takes this a step further, printing entire rocket bodies. At Plasmos, additive manufacturing allowed us to prototype components like engine nozzles and structural supports, cutting costs, and timelines while maintaining precision. Startups leveraging 3D printing can reduce dependency on traditional supply chains, a key advantage in the fast-paced world of aerospace.

Manufacturing in Space: The Orbital Economy

Manufacturing on Earth is being disrupted, but manufacturing in orbit promises to open entirely new frontiers. Companies like Varda Space and Redwire are pioneering zero-gravity manufacturing to produce materials that are stronger, lighter, and more precise than what is possible on Earth. At Plasmos, the potential of partnering with such orbital factories represented a unique opportunity to align in-space manufacturing with advanced delivery systems like the SpaceTruck. Entrepreneurs entering this field should focus on creating high-value products such as pharmaceuticals, fiber optics, or semiconductors that benefit from the microgravity environment.

Opportunities for Startups

The manufacturing revolution offers immense opportunities for new players in the aerospace ecosystem. Startups can innovate in areas such as:

- Designing specialized 3D-printed components for space and aviation.
- Developing modular orbital manufacturing units.
- Exploring advanced materials, such as alloys optimized for space environments.

These advancements are not just reshaping aerospace—they are laying the foundation for entirely new industries in orbit.

Artificial Intelligence (AI): The Brain of Aerospace

AI is no longer a luxury in aerospace; it is a necessity. As the industry grows more complex, AI plays an essential role in automating processes, improving efficiency, and driving innovation. Companies like SpaceX have been at the forefront, utilizing AI in both design and operations.

AI in Design and Manufacturing

AI is transforming how we design and build aerospace systems. SpaceX uses AI to optimize designs for its Starship, ensuring aerodynamic efficiency and structural integrity. Tools like Siemens NX and Dassault Systèmes CATIA leverage machine learning to refine designs with minimal human input. At Plasmos, AI was integral in modeling the complex thermodynamic properties of reusable rocket stages, enabling faster development cycles.

Startups like Relativity Space are pushing the boundaries further by combining AI with additive manufacturing. Their fully automated rocket factory integrates machine learning to optimize printing parameters in real time, reducing material waste while enhancing performance. These advancements demonstrate the power of AI as a catalyst for innovation.

Predictive Maintenance and Big Data Analytics

AI-powered predictive maintenance has revolutionized how aerospace systems are monitored and maintained. Platforms like Rolls-Royce's Intelligent Engine and Airbus' Skywise use AI to analyze real-time data, predicting failures before they occur. SpaceX employs similar techniques to monitor the health of its Falcon 9 rockets across multiple launches, ensuring reliability and safety.

For startups, predictive analytics offers significant opportunities. By developing tools that can analyze and predict system behavior, entrepreneurs can help operators minimize downtime and reduce costs. Combining big data and machine learning, startups can create niche solutions tailored to specific platforms or mission profiles.

AI in Operations and Automation

Beyond manufacturing and maintenance, AI has redefined flight operations. SpaceX's Starlink, powered by advanced AI algorithms, autonomously manages its satellite constellation to optimize global Internet coverage. Similarly, autonomous flight systems, such as Boeing's Autonomy on Demand, showcase how AI can streamline operations in crowded or complex environments.

For entrepreneurs, AI opens up possibilities in areas like autonomous navigation, mission planning, and even space traffic management as the number of satellites in orbit grows. By focusing on automation, startups can carve out significant roles in an increasingly interconnected aerospace ecosystem.

Technologies for Earth: Sustainable Flight and Energy

The need for sustainable and efficient transportation has never been more urgent. As urban populations grow and climate change accelerates, traditional aviation faces mounting pressure to reduce emissions and adapt to modern mobility needs. The shift from conventional aircraft like the Airbus A320 to electric vertical take-off and landing (eVTOL) vehicles and air taxis marks a profound transformation in aviation's role in our lives.

This evolution is not just about technological progress—it is about rethinking how we move people and goods, with a focus on sustainability, speed, and accessibility. But what comes next? The convergence of eVTOL, air taxis, supersonic travel, and space-based solar energy systems could define a future where efficiency and sustainability go hand in hand.

eVTOL: The Birth of Urban Air Mobility

Traditional aircraft like the Airbus A320 revolutionized commercial aviation, connecting distant cities and making air travel accessible to the masses. However, these planes were designed for long-haul and regional routes, often bypassing the growing needs of urban mobility. Enter eVTOL—a solution designed to bring aviation into our cities.

eVTOL vehicles represent a dramatic reimagining of air travel. They are smaller, quieter, and designed to take off and land vertically, eliminating the need for sprawling runways. Companies like Joby Aviation, Archer Aviation, and Volocopter are leading the charge, developing air taxis that promise to revolutionize urban transportation by alleviating traffic congestion and reducing commuting times.

But why is this shift so important? Cities like Los Angeles, New York, and Tokyo are already overwhelmed by traffic. Air taxis provide a solution by creating a new dimension of mobility—literally. Using distributed electric propulsion (DEP) systems, these aircraft are not only energy-efficient but also safer due to their multiple redundancies.

The Bigger Picture

While eVTOL addresses urban needs, the technology paves the way for a future where aviation is no longer constrained by traditional infrastructure. Imagine a world where supersonic aircraft like Boom Supersonic's Overture handle long-haul travel, while eVTOLs manage urban and regional mobility. This layered, integrated system could redefine how we think about air travel altogether.

Hydrogen-Powered Aviation: The Green Revolution for Long-Haul Flights

The transition from A320-like aircraft to hydrogen-powered aviation represents another critical leap. Hydrogen offers a zero-emission alternative to fossil fuels, addressing aviation's substantial carbon footprint. Airbus' ZEROe initiative envisions hydrogen-powered planes entering service by 2035, capable of connecting international cities without contributing to climate change.

Startups like Universal Hydrogen are innovating in this space, creating modular hydrogen capsules that can retrofit existing aircraft.

This approach accelerates adoption by bypassing the need for entirely new fleets. However, significant challenges remain, including the production of green hydrogen and the development of refueling infrastructure.

Hydrogen-powered aviation is not just an environmental imperative— it is an economic opportunity. As carbon taxes and regulatory pressures increase, airlines will need to adopt cleaner technologies to remain competitive. Startups that solve these bottlenecks—whether through improved fuel cells or scalable hydrogen production—stand to benefit immensely.

Harnessing Sunlight Energy from Space: The Next Frontier

Aerospace is not just reshaping mobility; it is revolutionizing energy. Space-based solar power is one of the most ambitious ideas to emerge from the industry, offering a solution to one of Earth's greatest energy challenges: how to generate renewable power continuously, even at night.

In my book *Electric Space: Space-Based Solar Power,* I explore how the unique conditions of space—such as uninterrupted sunlight and microgravity—make it an ideal environment for energy production. Companies like Space Solar Ltd. and SPS Alpha are developing technologies to collect solar energy in orbit and beam it back to Earth using microwaves or lasers. This technology has the potential to provide uninterrupted renewable energy to remote regions, stabilize energy grids, and reduce reliance on fossil fuels.

Why It Matters

Energy security is a global priority, especially as countries transition away from coal and natural gas. Space-based solar power could ensure a steady supply of clean energy, addressing not just environmental concerns but also geopolitical tensions tied to energy dependence.

For those interested in diving deeper into this revolutionary technology, *Electric Space* provides a comprehensive look at how space-based solar power could transform our planet's energy systems—and the startup opportunities it creates.

A Unified Vision: Air Taxis, Supersonics, and Sustainability

The integration of these technologies—eVTOL for urban mobility, hydrogen for long-haul flights, and space-based solar for energy—presents a compelling vision for the future of aviation. Imagine boarding an air taxi in your city, flying to a regional hub powered by hydrogen, and then connecting to a supersonic flight for intercontinental travel. All of this could be powered by solar energy collected in orbit.

This layered approach not only addresses sustainability but also enhances efficiency and convenience. Startups that focus on connecting these dots—developing technologies that integrate different modes of aviation or create synergies with energy systems—could play pivotal roles in shaping this future.

Opportunities for Startups

For entrepreneurs, the aerospace-to-Earth transition offers a wealth of opportunities:

- Innovating in battery technology to extend the range and efficiency of eVTOLs.
- Developing lightweight materials and fuel systems to make hydrogen-powered aviation scalable.
- Advancing space-based solar technologies to provide clean energy for aviation and beyond.
- Building software systems to integrate air taxis, supersonic flights, and renewable energy into a seamless global network.

The future of flight is closer than ever, and the technologies emerging today will define how we move and power our world tomorrow. For those bold enough to take the leap, this is the frontier where aerospace meets sustainability, and startups have the power to lead the charge.

Defense and Aerospace: War Technologies and Ethical Frontiers

The aerospace industry has long been tied to defense. Some of the greatest technological advancements in aviation, from jet engines to GPS, were driven by military needs. Today, the defense sector remains a significant driver of aerospace innovation, particularly in areas like unmanned systems, hypersonics, and autonomous decision-making. However, these advancements also bring new ethical challenges, especially as AI and automation blur the lines of accountability in warfare.

For startups, defense represents both opportunity and responsibility. It is a space where innovation can save lives, strengthen national security, and redefine the boundaries of technological capability—but it is also a domain that demands careful consideration of ethical implications.

Military Innovations: From Drones to Hypersonics

Drones have become the symbol of modern warfare, providing unparalleled precision and situational awareness. Iran's Shahed drones, for instance, are a prime example of how cost-effective unmanned aerial vehicles (UAVs) can achieve strategic objectives. These drones, which are both affordable and highly efficient, have influenced the global defense landscape, particularly in asymmetric warfare.

Startups like Anduril Industries are taking UAV innovation even further. Anduril integrates advanced AI and machine learning into its drone systems, enabling autonomous reconnaissance, surveillance, and even combat operations. Their Ghost drone, for example, uses onboard AI to analyze environments and make split-second decisions in the field, drastically reducing response times.

Beyond drones, hypersonic missiles and aircraft are pushing the boundaries of speed and precision. Companies like Hermeus and Raytheon are developing hypersonic systems that could change the balance of power in both defense and commercial applications. For startups, hypersonics offer opportunities in propulsion systems, thermal management, and advanced materials capable of withstanding extreme conditions.

Dual-Use Technologies: Where Defense Meets Civilian Applications

One of the most exciting trends in aerospace is the rise of dual-use technologies—innovations that serve both military and civilian purposes. For example:

- Autonomous drones developed for surveillance can be adapted for disaster response or agricultural monitoring.
- Hypersonic technologies designed for missiles can also enable faster cargo delivery or intercontinental travel.
- AI systems created for military decision-making can enhance civilian air traffic management.

This overlap creates immense opportunities for startups to address both markets. Plasmos Inc., for instance, focused on building versatile systems that could serve commercial satellite operators while aligning with defense needs. By designing dual-use technologies, startups can diversify revenue streams and maximize the impact of their innovations.

Ethics in Defense: The Responsibility of Innovation

While the potential of aerospace defense technologies is enormous, their ethical implications cannot be ignored. Autonomous weapons, for example, raise critical questions about accountability. If a drone powered by AI makes an incorrect decision in a combat zone, who is responsible—the developer, the operator, or the machine itself?

Programs like the Partnership on AI are working to establish ethical frameworks for the development and deployment of AI in defense. These frameworks emphasize the importance of human oversight, transparency, and accountability. Startups entering the defense sector must integrate these principles into their work to ensure public trust and long-term success.

Additionally, the use of UAVs in warfare has sparked debates about privacy, collateral damage, and the potential for escalation. It is crucial for startups to balance innovation with ethical responsibility, considering the broader consequences of their technologies.

Opportunities for Startups

For entrepreneurs, the defense sector offers immense potential for growth and innovation. Key areas of focus include:

- **Autonomous Systems:** Developing UAVs and other platforms that use AI to enhance decision-making, reduce human risk, and increase operational efficiency.

- **Counter-Drone Technologies:** As drones become more prevalent, so do the threats they pose. Startups can innovate in systems designed to detect and neutralize rogue drones.

- **Hypersonic Propulsion:** Designing propulsion systems and thermal protection technologies for hypersonic vehicles, which have both defense and commercial applications.

- **AI-Driven Decision Tools:** Building software that enhances situational awareness, mission planning, and operational execution, while incorporating ethical safeguards.

- **Cybersecurity:** Protecting critical aerospace systems from hacking and other digital threats, especially as AI and automation become more prevalent.

The Future of Aerospace Defense: Innovation with Accountability

The defense sector is evolving rapidly, and startups are at the forefront of this transformation. By leveraging advances in AI, autonomy, and materials science, small companies can deliver solutions that were once the domain of only the largest contractors. However, with this opportunity comes the responsibility to innovate ethically.

The future of aerospace defense lies in creating systems that not only perform better but also prioritize transparency, accountability, and the broader impact on society. For those who can balance these priorities, the defense sector offers unparalleled opportunities to lead, disrupt, and redefine what is possible.

The Future of Flight: Hypersonics and Suborbital Travel

For centuries, flight was limited to the realm of imagination, and even when it became a reality, it was constrained by speed and infrastructure. Today, however, the boundaries of what is possible in aviation are being redefined. Hypersonic travel and suborbital flights represent the next great leap in aerospace, promising to connect the world faster than ever before and extend human reach beyond the stratosphere.

These technologies are not just about speed; they are about creating a new way of experiencing the world. Imagine boarding a hypersonic plane in New York and landing in Tokyo in under 2 h, or taking a suborbital flight that gives you a panoramic view of Earth from space before dropping you off in another hemisphere. This is the future of flight—and startups are leading the charge.

Hypersonic Travel: Breaking the Speed Barrier Again

Hypersonic travel—defined as speeds exceeding Mach 5—is no longer confined to military applications. Companies like Hermeus, Boom Supersonic, and Reaction Engines are pushing the boundaries of what civilian air travel can achieve. Hermeus, for example, is developing the Quarterhorse, a hypersonic aircraft prototype designed to connect global cities in hours. Meanwhile, Boom Supersonics' Overture focuses on sustainable supersonic travel, bridging the gap between today's subsonic airliners and future hypersonic jets.

But why is hypersonic travel so significant? For one, it drastically reduces travel time, turning international trips into short commutes. Beyond the convenience, hypersonics also offer geopolitical and economic benefits by making the world more connected and accessible.

However, achieving hypersonic speeds comes with immense challenges. The heat generated at such velocities requires advanced thermal protection systems, while propulsion systems must be efficient enough to sustain speeds without excessive fuel consumption. Reaction Engines, a UK-based company, is addressing these challenges with its

SABRE engine, which combines jet and rocket engine technologies to enable sustained hypersonic flight.

Suborbital Travel: Redefining the Passenger Experience

While hypersonics focus on ultra-fast connections within Earth's atmosphere, suborbital travel takes passengers briefly into space before returning them to Earth. Companies like Blue Origin, Virgin Galactic, and SpaceX are pioneering this experience.

Blue Origin's New Shepard has already flown paying passengers to the edge of space, offering a few minutes of weightlessness and breathtaking views of Earth. Virgin Galactic's SpaceShipTwo provides a similar experience, focusing on making space tourism accessible to non-astronauts. These flights are not just about the destination—they are about the journey, giving people a taste of space and a new perspective on our planet.

Suborbital travel is not limited to tourism. SpaceX's Starship envisions a future where passengers and cargo can travel point-to-point across the globe via suborbital hops. Imagine boarding a Starship in Los Angeles and landing in Sydney in just over an hour. This kind of travel could revolutionize global logistics, making it possible to deliver critical supplies or humanitarian aid anywhere in the world almost instantly.

The Challenges and Opportunities of High-Speed Flight

While hypersonic and suborbital travel holds immense promise, they also come with significant technical and regulatory challenges. Startups in this space face hurdles such as:

- Thermal Management: Developing materials and systems to withstand extreme heat generated by hypersonic speeds.
- Fuel Efficiency: Creating propulsion systems that are both powerful and sustainable.
- Regulation: Navigating airspace laws, safety certifications, and environmental concerns.
- Passenger Experience: Designing cabins and systems that prioritize comfort, safety, and accessibility at extreme speeds or altitudes.

These challenges, however, present opportunities for startups to innovate. Companies can focus on developing advanced materials, such as lightweight alloys or composites, to reduce aircraft weight and increase durability. Others might work on improving environmental sustainability through hydrogen-powered or electric propulsion systems.

The Future: A Layered Transportation Network

The integration of hypersonic and suborbital travel into the global transportation system represents a vision of seamless connectivity. Hypersonic jets could serve major international routes, while suborbital flights could enable rapid point-to-point travel. Combined with eVTOLs and regional air mobility solutions, this layered network could redefine how people and goods move across the planet.

In this vision, startups play a critical role as innovators and disruptors. They bring agility, fresh perspectives, and the willingness to tackle the hardest problems—qualities that established aerospace giants often lack. By focusing on areas like propulsion, safety systems, and passenger experience, startups can lead the way in making high-speed flight a reality.

Opportunities for Startups

Entrepreneurs interested in high-speed aviation can explore opportunities in:

- Advanced Propulsion: Developing engines capable of sustaining hypersonic speeds or suborbital reentry.
- Thermal Protection: Innovating materials and cooling systems to manage the extreme heat generated at high speeds.
- Safety and Comfort: Designing cabins, seats, and systems optimized for passenger well-being during high-speed or suborbital travel.
- Logistics Applications: Creating solutions for cargo transportation that leverage the speed of hypersonics and the reach of suborbital systems.
- Integration with Current Systems: Building software and infrastructure to connect hypersonic and suborbital travel with traditional aviation and urban air mobility.

Why It Matters

Hypersonic and suborbital technologies have the potential to reshape not only how we travel but also how we connect as a global society. Faster travel reduces barriers between nations, enhances economic productivity, and provides new ways to respond to global challenges, from delivering aid to building stronger international ties.

For startups, this is more than just a technological opportunity—it is a chance to be part of a movement that redefines the human experience. By bringing bold ideas and innovative solutions to the table, entrepreneurs can help build a future where speed, sustainability, and connectivity converge.

Challenges and Ethics of Innovating

As aerospace technologies continue to push boundaries, they bring with them not only immense opportunities but also profound challenges. From environmental sustainability to ethical dilemmas surrounding AI and autonomous systems, the aerospace industry must confront critical questions about its impact on the world. Startups, as the new pioneers, have a unique responsibility to lead with innovation and integrity.

The following sections delve into the challenges that lie ahead and how entrepreneurs can address them while creating ethical, sustainable solutions.

Sustainability in Aerospace: Balancing Growth and Responsibility

Aerospace has traditionally been a resource-intensive industry, with a significant carbon footprint. The introduction of technologies like eVTOLs, hydrogen-powered aircraft, and space-based solar power represents a step toward sustainability, but the industry still faces major hurdles.

Environmental Challenges

- **Emissions:** Commercial aviation accounts for approximately 2.5% of global carbon dioxide (CO_2) emissions, with demand for air travel projected to grow.
- **Resource Use:** Manufacturing rockets, satellites, and aircraft requires energy-intensive processes and rare materials, contributing to environmental degradation.
- **Space Debris:** The growing number of satellites and space missions has resulted in over 27,000 pieces of trackable debris orbiting Earth, posing risks to both space operations and the environment.

Solutions and Opportunities

Startups can take a leadership role in developing sustainable practices, such as:

- **Clean Manufacturing:** Innovating in materials science to create lightweight, recyclable components that reduce resource consumption.
- **Circular Economy:** Adopting closed-loop systems where aerospace materials are reused or repurposed.
- **Debris Mitigation:** Developing technologies to clean up orbital debris or create self-deorbiting satellites.

Programs like Europe's Clean Sky 2 are already setting examples by funding research into sustainable aviation technologies. Startups that align with such initiatives can tap into funding, partnerships, and global recognition.

The Ethics of AI and Autonomous Systems

The rise of AI and automation in aerospace presents both exciting possibilities and daunting ethical challenges. Autonomous systems are revolutionizing areas like drone warfare, satellite operations, and flight navigation, but they also raise questions about accountability and transparency.

Ethical Dilemmas

- **Autonomous Decision-Making:** Who is responsible when an AI system makes a critical error, such as misidentifying a target in a combat zone or failing during an autonomous flight operation?

- **Bias in AI:** Algorithms trained on biased data could lead to unfair or harmful outcomes, particularly in applications like urban air mobility or defense.

- **Job Displacement:** Automation in manufacturing and operations may lead to significant job losses, creating societal and economic challenges.

Opportunities for Startups

- **Human–AI Collaboration:** Designing systems where AI augments human decision-making rather than replacing it entirely.

- **Transparent Algorithms:** Building AI models that are explainable and auditable, ensuring accountability in critical applications.

- **Ethical Oversight:** Startups can establish internal ethics boards or partner with organizations like the Partnership on AI to guide the responsible development of their technologies.

The aerospace industry needs ethical innovation to maintain public trust and ensure that advancements benefit society as a whole. Startups that embed these principles into their operations can differentiate themselves in an increasingly competitive market.

Social Impact: Access, Equity, and Global Implications

Technologies like eVTOLs, suborbital flights, and hypersonic travel promise to make the world more connected, but who will benefit? If left unchecked, these innovations could exacerbate existing inequalities, making advanced transportation accessible only to the wealthy or privileged regions.

The Problem

- **Urban Air Mobility:** Will air taxis cater only to affluent urban centers, leaving underserved communities behind?

- **Space Accessibility:** As private space travel becomes a reality, how do we ensure it does not become the exclusive domain of billionaires?

- **Global Disparities:** Developing nations often lack the infrastructure to adopt advanced aerospace technologies, widening the gap between the Global North and South.

The Solution
Startups can play a crucial role in democratizing access to aerospace technologies by:

- **Designing for Affordability:** Developing scalable, cost-effective solutions that can be adopted by a broader audience.
- **Collaborating with Governments:** Partnering with public agencies to subsidize infrastructure development in underserved regions.
- **Championing Inclusivity:** Ensuring that innovations address the needs of diverse populations, from urban commuters to remote rural communities.

Regulatory Challenges: Navigating a Complex Landscape
As aerospace innovation accelerates, regulatory frameworks are struggling to keep pace. Startups face a maze of international, national, and local regulations governing airspace, emissions, and safety standards.

Key Regulatory Hurdles
- **Airspace Management:** Integrating new technologies like drones and eVTOLs into crowded airspaces.
- **Sustainability Standards:** Complying with emerging environmental regulations, such as carbon emission caps and green certifications.
- **Safety Certification:** Meeting rigorous safety requirements for experimental technologies like hypersonic aircraft and autonomous systems.

Opportunities for Startups
- **Policy Advocacy:** Working with regulators to shape policies that balance innovation with safety and sustainability.
- **RegTech Solutions:** Developing tools to help aerospace companies navigate compliance requirements more efficiently.

- **Global Collaboration:** Participating in international efforts, such as the ICAO's Committee on Aviation Environmental Protection, to align innovations with global standards.

A Call to Innovate Responsibly

The challenges facing the aerospace industry are daunting, but they also present an opportunity for startups to lead with integrity. By prioritizing sustainability, ethics, and inclusivity, entrepreneurs can not only overcome these hurdles but also set a new standard for innovation.

Why It Matters

- Aerospace technologies have a profound impact on global connectivity, security, and quality of life.
- Addressing these challenges responsibly ensures that the benefits of innovation are shared widely and sustainably.
- Startups have the agility and creativity needed to tackle these problems head-on, shaping a future that balances progress with responsibility.

The future of aerospace is not just about building faster planes or reaching farther into space. It is about doing so in a way that respects our planet, prioritizes humanity, and creates a better world for all.

Conclusion: Your Moment to Build with Purpose

The aerospace industry is not just evolving; it is transforming—and so can you. When I look back at my journey, I see a series of leaps into the unknown. Each step, from building jet engines as a teenager in Tehran to founding Plasmos, was a gamble on the belief that the future could be different. That belief was not just about rockets or propulsion systems—it was about creating something meaningful and lasting.

This chapter is not just about me. It is about you—your dreams, your ideas, and your potential to disrupt an industry that needs bold thinkers more than ever.

The Courage to Start

I did not have a blueprint when I started Plasmos. The idea of building a reusable third-stage rocket with hybrid propulsion was not something people were lining up to fund. But I believed in it. I believed in the possibility of making space more accessible, and I knew the only way to turn that belief into reality was to take the first step—no matter how daunting.

Founding Plasmos meant late nights, countless rejections, and moments of doubt. It also meant finding incredible teammates, raising $555,000 from Velo3D and Plug and Play Ventures, and pitching to decision-makers in the Pentagon and the White House. Each small victory built momentum, proving that even the most audacious dreams are achievable with persistence and grit.

LISA and the Art of the Pivot

Before Plasmos, there was LISA Deutschland GmbH—a startup I founded during one of the most uncertain times in recent history. The aviation industry was crumbling under the weight of the COVID-19 pandemic. We could have folded, but instead, we pivoted. We shifted from predictive maintenance to developing a defense data lake platform and secured contracts with Airbus Defense and Space. That pivot not only saved the company but also led to its acquisition by Helios GmbH.

LISA taught me that resilience is not just a trait—it is a strategy. Startups are about adapting, learning, and finding opportunities in adversity. If you are thinking about starting your own company, remember: failure is not the opposite of success. It is part of the journey.

Why This Moment Matters

The aerospace industry is at an inflection point. Technologies like eVTOLs, hypersonic travel, and space-based solar power are reshaping how we think about mobility and energy. But the most exciting part? These innovations are not coming from government programs or massive corporations alone—they are coming from startups.

Startups like SpaceX redefined what was possible in space exploration. Relativity Space is printing entire rockets. And Plasmos? We challenged the conventional wisdom about reusability and cost in space mobility.

This moment in history belongs to those willing to think big, take risks, and redefine the rules.

A Personal Reflection: Why I Build

When I was 15, I built a jet engine—not because anyone asked me to, but because I was fascinated by what I could create with my own hands. That curiosity has guided me ever since. Whether it was managing a $400 million risk portfolio at Airbus or founding Plasmos, I have always been driven by the question: What can we build that will make the world better?

But building is not just about technology. It is about people. The most rewarding part of my journey has been working with brilliant, passionate individuals who share a vision for what is possible. At Plasmos, we were not just designing rockets—we were building a culture of innovation, resilience, and purpose.

Your Call to Action

To every student or aspiring entrepreneur reading this: the aerospace industry needs you. Not tomorrow. Not someday. Now. The challenges we face—climate change, energy security, global connectivity—will not wait. And neither should you.

If you are sitting in a classroom right now, dreaming about the next big thing, I want you to ask yourself: What is stopping me? The truth is you do not need permission to start. You do not need all the answers, either. All you need is the courage to take the first step.

What I Wish I Knew

If I could go back and give my younger self advice, it would be this:

- **Embrace the Unknown:** You do not need to have everything figured out to start. Uncertainty is where growth happens.
- **Find Your Tribe:** Surround yourself with people who believe in your vision and push you to be better.
- **Think Beyond Profit:** Build something that solves real problems and makes a lasting impact.
- **Fail Forward:** Every setback is an opportunity to learn. Use it.

The Future Is Yours to Build

The sky is not the limit—it is the launchpad. The future of aerospace is not being written by people who play it safe. It is being written by those who dare to dream, build, and disrupt. Whether you are creating the next breakthrough in propulsion, designing sustainable energy systems, or revolutionizing urban air mobility, know that your work matters.

So, to the dreamers, the builders, and the risk-takers: Take the leap. Start your company. Build your rocket. Change the world. The future is waiting—and it belongs to you.

Glossary

Ablation - The process by which heat is removed from a reentry body through the melting, vaporizing, and scouring off of special materials as the body is heated by friction with the atmosphere.

Antenna - Structure used to receive and send electromagnetic waves. It adapts the system and free space impedance.

Apogee - In an orbit, the point farthest away from the Earth.

Array - Process of combining more than one antenna to mathematically synthesize a bigger equivalent antenna.

Command - Data sent from the Earth to the spacecraft.

Doppler data - Data extracted from the carrier signal. It is used to measure the spacecraft radial velocity.

Free space losses - Signal attenuation due to the spherical shape of an electromagnetic wave as it propagates.

Gimbal - A pivot device that allows the entire nozzle and the flow of exhaust gases to be swiveled in any desired direction, thus directing the rocket's course in the opposite direction.

Ground station - The Earth facility used to send and receive data to and from the spacecraft.

Modulation - The process of getting information into carrier electromagnetic waves. We can modulate wave amplitude, frequency, and phase.

Parking orbit - The initial orbit achieved for a vehicle that is to be sent into space as a probe, for vehicles to be joined in rendezvous, or for reentry vehicles to be returned to Earth.

Payload - The load carried by a spacecraft consisting of the things (such as instruments or satellites) necessary for the purpose of flight.

Perigee - In an orbit, the point closest to the Earth.

Polar orbit - Orbit that passes over Earth's poles. Its inclination is 90°. It allows the satellite to cover different areas of the planet.

Propellant - The solid or liquid fuel of a rocket that, when burned in combination with an oxidizer, provides thrust.

Radiation pattern - The diagram showing how an antenna radiates in space.

Ranging signal - Signal used to obtain the exact radial distance to the spacecraft.

Satellite - Anything held in an orbit by the gravitational field of another body.

Stage - One unit in a multistage rocket. The stage has its own engines and fuel supply and may have its own guidance system. The booster is the first stage of a multistage rocket.

Telemetry - Data sent by spacecraft to Earth.

Thrust - The amount of push generated by a rocket engine measured in kilograms.

Bibliography

3D Printing Media Network, "GE9X, the Largest and Most 3D Printed Jet Engine Ever, Is Flying," 2018, Web, March 30, 2019.

Aerospace Assuring Space Mission Success, "A Brief History of Space Exploration," Web, May–June 2014, http://www.aerospace.org/education/inspiring-the-next-generation/space-primer/a-brief-history-of-space-exploration/.

Amos, J., "Airbus Drops Model 'Space Jet'," BBC News, BBC, June 3, 2014, Web, December 18, 2014.

Anderson, J.D., "Aerodynamics, Flight," in *Introduction to Flight* (New York: McGraw-Hill, 1989).

Associated Newspapers, "Pictured: Rocketman Flies over Alps with Jet-pack Strapped to His Back," Mail Online, May 2008, Web, January 7, 2015.

Belfiore, M., "Under Construction: XCOR's Space Corvette," Popular Mechanics, August 13, 2012, Web, May 1, 2014.

Bellis, M., "The Dynamics of Airplane Flight," About.com Inventors, About.com, March 5, 2014, Web, February 22, 2015.

Dan, S., "What Is a Satellite?," Institute for Global Environmental Strategies, NASA, February 2014, Web, Autumn 2014, http://www.nasa.gov/audience/forstudents/5-8/features/what-is-a-satellite-58.html.

Doody, D. and Fisher, D., "A Gravity Assist Primer: Mechanical Simulator," NASA Jet Propulsion Labratory, Web, April–May 2014.

Ducksters, "Physics for Kids," Basics of Sound, Ducksters Education Site, Fall 2014, http://www.ducksters.com/science/sound101.php.

Dunbar, B., "Mars One Project," Ideas Based on What We'd Like to Achieve, NASA Glenn Research Center, November 23, 2004.

Dunbar, B., "Ion Propulsion," NASA, May 21, 2008, http://www.nasa.gov/centers/glenn/about/fs21grc_prt.htm.

Dunbar, B., "How Do Planes Fly?," NASA, May 3, 2010.

Dunbar, B., "Sometimes Size DOES Matter: 25 Years with the Largest Wind Tunnel in the World," NASA, December 14, 2012.

ESA, "Electric Propulsion," 2014, http://www.esa.int/esapub/br/br187/br187.pdf.

Hitt, D., "What's an Orbit?," NASA Educational Technology Services, NASA, July 2014.

IHS Engineering 360, "Chapter 2: A Review of Basic Propulsion," On GlobalSpec, January 7, 2015.

Jones, D.R. and Baghchehsara, A., *Electric Space: Space-Based Solar Power Technologies & Applications: Unlimited Energy from Space* (North Charleston, SC: CreateSpace, 2013).

Kellner, T., "The Blade Runners: This Factory Is 3D Printing Turbine Parts for the World's Largest Jet Engine," GE Reports, 2018, Web, March 29, 2019.

Kraus, J.D., *Antennas* (New York: McGraw-Hill, 1988).

Lunar and Planetary Institute, "Space Exploration—The Next Generation," Humans in Space Art, Web, June–July 2014.

Maral, G. and Bousquet, M., *Satellite Communications Systems: Systems, Techniques, and Technology* (Hoboken, NJ: John Wiley & Sons, 1993).

MiGFlug & Adventure GmbH, "Xcor Lynx Space Flight: The Age of Space Flights Is about to Begin!," Fly Fighter Jet Dot Com, June–July 2014.

NASA Glenn Research Center, "Beginner's Guide to Rockets," 2014, http://exploration.grc.nasa.gov/education/rocket/.

NASA Glenn Research Center, "Ultra-Efficient Engine Technology," Engines, Web, Fall 2014.

NASA-USA.DE, "What Is Orion," n.d., http://www.nasa-usa.de/audience/forstudents/5-8/features/nasa-knows/what-is-orion-58.html.

Physics Tutorial, "The Physics Classroom Online," Kepler's Three Laws, Web, June–July 2014, http://www.physicsclassroom.com/class/circles/Lesson-4/Kepler-s-Three-Laws.

Proakis, J.G., *Digital Communications* (New York: McGraw-Hill, 1983).

Ross, Y., "Jetman Dubai," Yves Ross, March 1, 2014.

Science Buddies, "How to Build and Use a Subsonic Wind Tunnel," 2014.

Science Kids, "Fun Flight Facts for Kids—How Planes Fly, Wright Brothers, Aviation Information," Science Kids, NZ, July 24, 2014.

Scientific Figure on ResearchGate, "Space as a Tool for Astrobiology: Review and Recommendations for Experimentations in Earth Orbit and Beyond," accessed March 31, 2019, https://www.researchgate.net/figure/Principal-orbits-for-artificial-satellites-around-the-Earth_fig34_318029811.

Smithsonian Education, "How Things Fly. Smithsonian National Air and Space Museum," 2014.

Space.com, "Elon Musk Unviels Spacex Mars Colony Ship," n.d., http://www.space.com/34210-elon-musk-unveils-spacex-mars-colony-ship.html.

Stillman, D., "What Is a Satellite?," Institute for Global Environmental Strategies, NASA, 2014, http://www.nasa.gov/audience/forstudents/5-8/features/what-is-a-satellite-58.html.

Teachertech, "Newton's 3 Laws of Motion," www.Teachertech.rice.edu, 2019, Web, March 31, 2019.

The Physics Classroom, "Kepler's Three Laws," Physics Tutorial, June–July 2014, http://www.physicsclassroom.com/class/circles/Lesson-4/Kepler-s-Three-Laws.

Voskuijl, M., "Introduction to Aeronautical Engineering," Flight Mechanics 1-4, TU Delft, September–October 2014.

Wikimedia Foundation, "Flight Control Surfaces," Wikipedia, Web, Summer 2014.

Wikipedia, "Newton's Laws of Motion," Web, February 11, 2015.

Woodford, C., "Helicopters," How Does a Helicopter Work?, Explainthatstuff, August 2, 2014, Web, May–June 2014.

Woollaston, V., "'Budget' XCOR Space Trip Set to Launch in 2016 Will Let You Pilot the Ship for £57,000," Mail Online, Associated Newspapers, December 24, 2014.

Yechout, T.R. and Morris, S.L., *Introduction to Aircraft Flight Mechanics: Performance, Static Stability, Dynamic Stability, and Classical Feedback Control* (Reston, VA: American Institute of Aeronautics and Astronautics, 2003).

Index

About the Author

Ali Baghchehsara

Ali Baghchehsara is a distinguished entrepreneur, aerospace engineer, and investor renowned for pioneering solutions in aerospace, AI, and cognitive systems. His contributions have earned him the Royal Aeronautical Society's Achievement Award (2014), and the American Society of Mechanical Engineers' Outstanding Service Award (2015). In

Courtesy of Ali Baghchehsara.

2013, he co-authored *Electric Space: Space-based Solar Power Technologies & Applications* with Danny Royce Jones Sr., published in North Charleston, SC, introducing innovative concepts for satellite-based solar energy. At Airbus, he pioneered the first Autonomous System for Distress Tracking (ADT) for commercial aircraft, setting a benchmark in autonomous aviation.

With a Bachelor's degree in Aerospace Engineering and a Master's degree in Aeronautics and Management, Ali has authored 12 books, including the Amazon best-seller *Fundamentals of Aerospace Engineering: (Beginner's Guide)*, which demystifies aerospace for diverse readers, and *Physical Mechanism of Love*, exploring human relationships through physics. He founded LISA Group, delivering predictive maintenance solutions, which he sold in 2021. Currently, he leads a stealth startup advancing AI and aerospace innovations. As founder of Plasmos, he develops dual-mode propulsion systems for space vehicles, leveraging electric and chemical technologies. A Forbes Technology Council member, Ali continues to shape industries through visionary leadership, investment, and digital transformation.

About the Editors

Francisco Gallardo López

Francisco Gallardo López was born on June 6, 1986, in Madrid, Spain. As a talented radio frequency engineer, he obtained his telecommunication degree specializing in radio frequency from the Polytechnic University of Madrid. Francisco earned both the ROHDE & SCHWARZ Price for the best final degree project in radio frequency and the final degree projects and academic records national contest "Liberalización de las Telecomunicaciones" from the Spanish National Official College of Telecommunications Engineers (COITT). In 2012, he joined NASA's Madrid Deep Space Communication Complex as a systems engineer. Francisco is a creative person with innovative approaches to aerospace development in the area of radio frequency design with deep knowledge of ground stations, radio frequency, and aerospace.

Courtesy of Francisco Gallardo López.

Jens H. Strahmann

Jens H. Strahmann has more than 35 years' experience in the aerospace arena and has worked for civil and military programs including fighters, unmanned aerial vehicles (UAVs), passenger, transport, conversion aircrafts, and airships. In these special environments, flight test instrumentation has been one of his focal points. Instrumenting prototypes for applying the airwor-

thiness certification and deep discussions with test pilots, flight test engineers, program management, and the certification authorities European Union Aviation Safety Agency (EASA) and Federal Aviation Administration (FAA) have provided him a complete picture of the complete certification process. As a matter of fact, being involved in development programs and being responsible, depending on the program, for the final assembly line and integration of airships, pre-integration of aircrafts, lab tests, iron birds, and ground and flight tests for aircraft systems, enables a person to judge the pros and cons of the different organizations. The international character of all the different prototype aircrafts has widened his view of the collaboration necessary to make a program a success. In the end, people are behind all the different tasks and they are making a difference.

One of the highlights of his career has been the development, integration responsibility, and successful flight test campaign of the fly-by-wire cabinet system in the frame of the German-founded Technology program electronic flight control system (EFCS) on VFW 614.

After integration and founding the single company Airbus, he was responsible for flight test integration and flight test for passenger-to-freighter conversions A310 and tanker.

Later, he was responsible for the creation and operation of the certification Test Center for all High Lift Systems on Airbus aircrafts (A320 family, A330, A340, A380, A400M, and A350).

Edward G. Gibson, PhD

Edward G. Gibson, PhD, was born on November 8, 1936, in Buffalo, New York (NY). Dr. Gibson was selected as a scientist–astronaut by NASA in June 1965. He completed a 53-week course in flight training at Williams Air Force Base, Arizona, and earned his Air Force wings. Since then, he has flown helicopters and the T-38. He served as a member of the astronaut support crew and as a capcom for the Apollo 12 lunar landing. He has also participated in the design and testing of many elements of the Skylab space station.

Courtesy of Edward G. Gibson.

Dr. Gibson was the science pilot of Skylab 4 (third and final manned visit to the Skylab space station), launched on November 16, 1973, and concluded on February 8, 1974. This was the longest manned flight (84 days 1 h 15 min) in the history of manned space exploration to date. Dr. Gibson was accompanied on the record-setting 34.5-million-mile flight by Gerald P. Carr (commander) and William R. Pogue (pilot). They successfully completed 56 experiments, 26 science demonstrations, 15 subsystem detailed objectives, and 13 student investigations during their 1214 revolutions of the Earth. They also acquired a wide variety of the Earth resources observation data using Skylab's Earth resources experiment package camera and sensor array. Dr. Gibson was the crewman primarily responsible for the 338 h of Apollo Telescope Mount operation, which made extensive observations of solar processes.

Until March 1978, Dr. Gibson and his Skylab 4 teammates held the world record for individual time in space: 2017 h 15 min 32 sec, and Dr. Gibson logged 15 h and 17 min in three extravehicular activities (EVAs) outside the orbital workshop. Recreational interests include distance running, swimming, photography, flying, and motorcycling. He graduated with a doctorate in engineering with a minor in physics from the California Institute of Technology in June 1964; an honorary doctorate of science from the University of Rochester (NY) in 1974; and

an honorary doctorate of science from Wagner College, Staten Island, NY, in 1974.

He is elected a fellow of the American Astronautical Society and a member of the American Institute of Aeronautics and Astronautics, Tau Beta Pi, Sigma Xi, and Theta Chi. He received the JSC Certificate of Commendation (1970), the NASA Distinguished Service Medal (presented by President Richard M. Nixon in 1974), the City of New York Gold Medal (1974), the Robert J. Collier Trophy for 1973 (1974), the Dr. Robert H. Goddard Memorial Trophy for 1975 (1975), the Federation Aeronautique Internationale's De La Vaulx Medal and V. M. Komarov Diploma for 1974 (1975), the American Astronautical Society's 1975 Flight Achievement Award (1976), the AIAA Haley Astronautics Award for 1975 (1976), a Senior U.S. Scientist Award from the Alexander von Humboldt Foundation (1976), and a JSC Special Achievement Award (1978).

While studying at Caltech, Gibson was a research assistant in the field of jet propulsion and classical physics. His technical publications are in the fields of plasma physics and solar physics. He was a senior research scientist with the Applied Research Laboratories of Philco Corporation at Newport Beach, California, from June 1964 until coming to NASA. While at Philco, he did research on lasers and the optical breakdown of gases. Subsequent to joining NASA in 1965, he wrote a textbook in solar physics entitled *The Quiet Sun*. Gibson's training and data acquisition as a science pilot on the last Skylab mission were in the areas of solar physics, comet observations, stellar observations, the Earth resources studies, space medicine and physiology, and flight surgeon activities. He has logged more than 4300 h flying time—2270 h in jet aircraft.

Ramtin Jamshidi

Ramtin Jamshidi was born in 1987 in Tehran, Iran. He obtained a Bachelor of Visual Communication degree from the University of Art in Tehran and a Master of Communication Design from Swinburne University of Technology in Melbourne, Australia. Since 2009, he has been an active freelance designer doing many projects in both Iran and Australia. Ramtin is a talented young designer who has been selected in many international poster contests around the world.

Courtesy of Ramtin Jamshidi.

www.ingramcontent.com/pod-product-compliance
Lightning Source LLC
Chambersburg PA
CBHW041159220326
41597CB00001BA/7